William A. Foley
A Sketch Grammar of Kopar

Pacific Linguistics

Managing editor
Alexander Adelaar

Editorial board members
Wayan Arka
Danielle Barth
Don Daniels
T. Mark Ellison
Bethwyn Evans
Nicholas Evans
Gwendolyn Hyslop
David Nash
Bruno Olsson
Bill Palmer
Andrew Pawley
Malcolm Ross
Dineke Schokkin
Jane Simpson

Volume 667

William A. Foley
A Sketch Grammar of Kopar

A Language of New Guinea

DE GRUYTER
MOUTON

ISBN 978-3-11-135754-6
e-ISBN (PDF) 978-3-11-079144-0
e-ISBN (EPUB) 978-3-11-079154-9
ISSN 1448-8310

Library of Congress Control Number: 2022933736

Bibliographic information published by the Deutsche Nationalbibliothek
The Deutsche Nationalbibliothek lists this publication in the Deutsche Nationalbibliografie;
detailed bibliographic data are available on the internet at http://dnb.dnb.de.

© 2023 Walter de Gruyter GmbH, Berlin/Boston
This volume is text- and page-identical with the hardback published in 2022.
Cover image: David Forman/Photodisc/Getty Images
Typesetting: Integra Software Services Pvt. Ltd.
Printing and binding: CPI books GmbH, Leck

www.degruyter.com

Preface

I would like to dedicate this work to the people of Wongan and especially Kopar villages for generously sharing their marvelous language with me. While I fully understand their very valid reasons for abandoning their traditional village language for the lingua franca of Papua New Guinea, Tok Pisin, I cannot help but lament the loss of such a magnificent product of human cultural endeavor that is the Kopar language. Languages, and particularly polysynthetic languages, are in retreat everywhere, and with these losses vanish unique mirrors into the amazing potentials of human linguistic creativity. Some hold that languages are fundamentally much the same everywhere. I reject such a view. While there are some common constraints that hold because of human brain neurophysiology, variation and diversity across languages is very real, and Kopar provides a view into some of what is possible in its impressive compression of concepts into single words. We are moving inexorably into a world of linguistic monoculturalism, and while there are some benefits in ease of wider communication, there are incalculable losses in the incredible diversity of expression that the human linguistic capacity allows. It is in the service of documenting some of the richness of that capacity as exemplified in Kopar that I have written this monograph.

Contents

Preface —— V

List of Abbreviations —— XI

List of Map —— XIII

List of Figures —— XV

List of Tables —— XVII

Chapter 1
Introduction —— 1

Chapter 2
Phonology —— 6
2.1 Phonemes and allophones —— 6
2.2 Phonotactics —— 8
2.3 Stress —— 13
2.4 Morphophonology —— 13
2.5 A note on presentation of examples —— 22

Chapter 3
Word classes —— 24
3.1 Nouns —— 25
3.2 Verbs —— 26
3.3 Adjectives —— 30
3.4 Quantifiers —— 36
3.5 Pronouns —— 39
3.6 Deictics —— 41
3.7 Postpositions —— 46
3.8 Temporals —— 49
3.9 Conjunctions —— 51
3.10 Interjections —— 53

Chapter 4
Nouns and noun phrases —— 54
4.1 Nouns —— 54
4.2 Noun compounds —— 57

4.3	Noun phrases —— 57
4.3.1	Possession in noun phrases —— 57
4.3.2	Postnominal modification in noun phrases —— 59
4.4	Relative clauses —— 60

Chapter 5
Verbal morphology —— 65

5.1	Transitivity —— 66
5.2	The pronominal affix agreement systems for core arguments —— 68
5.2.1	The accusatively aligned system —— 68
5.2.2	The ergatively aligned system —— 71
5.2.2.1	Neutral inflectional pattern: Non-local person acts on non-local person —— 75
5.2.2.2	Direct inflectional pattern: Local person acts on non-local person —— 77
5.2.2.3	Inverse inflectional pattern: Non-local person acts on local person —— 79
5.2.2.4	Inverse inflectional pattern: Local second person acts on local first person —— 81
5.2.2.5	Impersonalization: Local first person acts on local second person —— 83
5.2.3	Pronominal agreement in the perfective —— 86
5.2.4	The dative suffixes —— 90
5.3	Tense, aspect and mood —— 96
5.3.1	Tense —— 97
5.3.1.1	The far past tense —— 99
5.3.1.2	The near past tense —— 99
5.3.1.3	The present tense —— 99
5.3.1.4	The perfective aspect —— 101
5.3.1.5	The immediate future tense —— 102
5.3.1.6	The future tense —— 103
5.3.2	Aspect —— 104
5.3.2.1	Progressive aspect —— 104
5.3.2.2	Durative aspect —— 105
5.3.2.3	Extended aspect —— 106
5.3.3	Modality —— 107
5.3.3.1	Negation —— 107
5.3.3.2	Ability —— 109
5.3.3.3	Necessity or obligation —— 110

5.3.3.4	Permissive modality —— 111	
5.3.4	Mood or illocutionary force —— 112	
5.3.4.1	Questions —— 112	
5.3.4.2	Imperatives —— 115	
5.3.4.3	Prohibitives —— 117	
5.3.4.4	Hortatives —— 119	
5.4	Verb stem derivations: Valence changes —— 120	
5.4.1	From transitive verb root to intransitive verb stem —— 121	
5.4.1.1	Reflexivization —— 121	
5.4.1.2	The detransitivizer —— 123	
5.4.2	From intransitive verb root to transitive verb stem —— 125	
5.4.2.1	Causatives —— 125	
5.4.2.2	Applicatives —— 127	
5.4.2.2.1	The comitative applicative —— 127	
5.4.2.2.2	The general applicative —— 129	
5.5	Verb theme derivations —— 132	
5.5.1	Possessor raising —— 133	
5.5.2	Incorporation —— 135	
5.5.2.1	Incorporation of temporals —— 136	
5.5.2.2	Incorporation of adverbials —— 138	
5.5.2.2.1	*nda-* 'now' —— 139	
5.5.2.2.2	*ŋga-* ~ *ka-* 'first' —— 141	
5.5.2.2.3	*pa-* 'still, yet' —— 141	
5.5.2.2.4	*mbi-* 'again' —— 142	
5.5.2.3	Incorporation of directionals —— 143	
5.5.2.4	Incorporation of nouns —— 145	
5.5.2.5	Incorporation of verbs: Verb serialization —— 145	

Chapter 6
Clause structure —— 151

6.1	Basic verbal clauses —— 151
6.1.1	Constituent order in verbal clauses —— 151
6.1.2	The expression of grammatical functions in verbal clauses —— 154
6.1.2.1	Grammatical relations in intransitive and transitive clauses —— 154
6.1.2.2	Verbs with cognate objects —— 156
6.1.2.3	Dative arguments and ditransitive clauses —— 157
6.1.2.4	Experiential clauses —— 162
6.1.3	Marking of oblique roles by postpositions —— 163

6.1.3.1	The dative postposition *ŋga* —— 163	
6.1.3.2	The purposive postposition *ndək* —— 166	
6.1.3.3	The source postposition *ta(r) ~ tar o ndək* —— 167	
6.1.3.4	Locative postpositions —— 168	
6.1.3.5	The comitative postposition *nda* —— 170	
6.1.3.6	The oblique case marker *-mb* —— 171	
6.2	Nonverbal clauses —— 173	
6.2.1	Nonverbal clauses of identification —— 173	
6.2.2	Nonverbal clauses of location —— 174	
6.2.3	Nonverbal clauses of attribution —— 175	
6.2.4	Nonverbal clauses of possession —— 176	

Chapter 7
Interclausal relations —— 180

7.1	Non-finite constructions —— 180
7.1.1	Dative infinitive constructions —— 180
7.1.2	Purposive infinitive constructions —— 185
7.1.3	Nominalization constructions —— 186
7.2	Finite constructions —— 187
7.2.1	Finite subordinate clauses —— 187
7.2.2	Coordination of full independent clauses —— 190
7.2.3	Clause chaining —— 193

Appendix 1 —— 205

Appendix 2 —— 211

References —— 243

Index —— 245

List of Abbreviations

ABIL	ability modality
ADV	adverbial
APPL	general applicative
C	consonant
CAUS	causative
COM	comitative
COUNTERFACT	counterfactual
DAT	dative
DEF	definite
DEP	dependent
DES	desiderative
DETR	detransitivizer
DIST	distal deictic
DL	dual number
DUR	durative aspect
EMPH	emphatic
ERG	ergative
EXT	extended aspect
FR.PAST	far past tense
FUT	future tense
IM.FUT	immediate future tense
IMP	imperative mood
INDEF	indefinite
INV	inverse
IRR	irrealis
ITR	intransitive
LK	linking phoneme
LOC	locative
NE	derives forms like *mine, thine* (4.3.1)
NEC	necessary modality
NEG	negative
NMLZ	nominalizer
NR.PAST	near past tense
OBJ	direct object
OBL	oblique
PC	paucal number
PERM	permissive modality
PFV	perfective aspect
PL	plural number
PRES	present tense
PROG	progressive aspect
PROHIB	prohibitive mood
PROX	proximate deictic
PURP	purposive

Q	question
RED	reduplication
REFL	reflexive
SEQ	sequential
SG	singular number
TR	transitive
V	vowel
wh	wh-element
1	first person
2	second person
3	third person

List of Map

Map 1 Distribution of the Lower Sepik languages —— 3

List of Figures

Figure 1 The Lower Sepik language family —— 1
Figure 2 Kopar consonantal phonemes —— 6
Figure 3 Kopar vowel phonemes —— 7
Figure 4 The temporal continuum in Kopar —— 98

List of Tables

Table 1	Kopar independent pronouns —— 39	
Table 2	Kopar possessive pronouns —— 58	
Table 3	Bound subject pronominals in unrealized tense-moods —— 69	
Table 4	Bound subject (nominative) pronominals for realized tenses in intransitive verbs —— 71	
Table 5	Analysis of Kopar pronominal prefixes for intransitive verbs in realized tenses —— 72	
Table 6	Bound subject (ergative) pronominals for realized tenses in transitive verbs —— 74	
Table 7	Bound pronominal affixes (nominative) in the perfective for intransitive verbs —— 86	
Table 8	Bound pronominal affixes in the perfective for transitive verbs —— 88	
Table 9	Dative suffixes —— 91	
Table 10	Bound pronominal suffixes for imperative mood —— 115	
Table 11	Incorporated temporal suffixes —— 136	
Table 12	Incorporated directional suffixes —— 143	
Table 13	Inflection of the verb of possession —— 179	

Chapter 1
Introduction

Kopar is a now moribund language formerly spoken in three villages at or near the mouth of the Sepik River in the northern swampy lowlands of Papua New Guinea. Kopar is one of the six languages of the Lower Sepik family within the larger Lower Sepik-Ramu family (Foley 2017a), and is most closely related to the language to its immediate west, Murik, with which it forms a sub-group (see the comparative wordlist in Appendix 1). Figure 1 presents the languages of the Lower Sepik family and their relationships, a language family whose internal diversity is roughly on the order of Germanic:

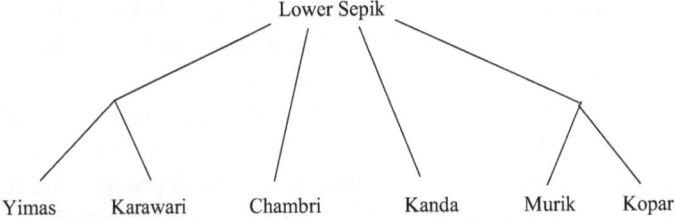

Figure 1: The Lower Sepik language family.

Kopar was spoken in three villages, Kopar (3.862426°S, 144.525852°E), located right at the mouth of the Sepik River, Singrin (3.939457°S, 144.430355°E), located about thirty kilometers upriver and Wongan (3.999326°S, 144.532123°E), found in the mangrove swamps to the southeast of Kopar village, perhaps ten kilometers as the crow flies. I write 'was spoken' because the language was already moribund twenty-five years ago when I and an honors student of mine, Stephen Hill, did our fieldwork and hardly used in daily life. Language shift was already very far advanced in the mid 1990s, with Tok Pisin already having nearly entirely usurped the functions of Kopar. It appears that large scale language shift already started to occur in the 1960s in Kopar, about ten to twenty years earlier than elsewhere in the lower and middle Sepik River regions. I suspect very few fluent speakers remain today, perhaps less than two dozen across the three villages, although that is an unsure estimate, and even they would rarely use the language. Certainly, use of the language has by now even more retreated in daily life.

The moribund state of the language even twenty-five years ago created problems in fieldwork similar to language salvage work. Speakers were sometimes unsure of correct forms in the complex verbal morphology and gave conflicting forms when they were elicited, although multiple checking usually allowed a consensus to be arrived at. When possible, elicited forms have always been checked against the spontaneous forms provided in the four narrative texts for accuracy. It is also possible, of course, that there was extensive variation in forms employed across speakers even in such a small speech community, but the very limited time span of fieldwork prevented any serious study of language variation, either idiosyncratic or gender or age based. Because of these difficulties, in this short grammar I have identified areas where analyses need to be regarded as provisional, pending further data should their collection ever prove to be possible. Furthermore, the whole amount of fieldwork time devoted to Kopar was only around a month, as this was a side project to my main research on the documentation of the neighboring distantly related Watam language, so data on the language are limited in any case. The data were almost all collected in Kopar village, with a small amount in Wongan village; in addition, I have a word list of some hundred items (Abbott 1985) collected in the Singrin dialect. There is some, relatively minor, dialect differentiation between the three villages. This description is based on the dialect of Kopar village, but I will occasionally comment on differences in the Wongan dialect. The data consist of word lists, nominal paradigms, verbal paradigms across tenses and moods for intransitive and transitive verbs, basic clause permutations and three narrative texts of moderate length and one shorter narrative text. Recordings of all field materials have been deposited with PARADISEC: https://catalog.paradisec.org.au/collections/WF1. Generally, a grammatical description would require more than this. And a month is certainly insufficient to give a full description of any language, and this especially holds true of one so complex and morphologically rich as Kopar. But given the quite unusual and striking typological features of the language, its importance in reconstructing the comparative grammar of the Lower Sepik family, and the fact that further documentation is unlikely to occur or sadly even be possible, I offer this sketch grammar to the linguistics research community and especially to Papuanists. Unfortunately given the sparse data, there will be gaps in the description offered here and almost certainly some errors and overgeneralizations. Like all Lower Sepik languages Kopar is very morphologically complex (see Foley (1991, 2017b) on Yimas), and there is no way this short grammar nor the limited data collected can do justice to that. In particular, some of the rich incorporating and derivational morphology of the verb undoubtedly remains undocumented or underanalyzed. For instance, *kawari-* 'bury' and *kawarumbut-* 'throw down from a height' are clearly derivationally related, but the data are insufficient to ascertain their morphological structure. Similarly, *niŋja-* 'send', *paneŋja-* 'aim to thrust', *ruruŋja* 'shake' and *təmeŋja-* 'tell' are derived from *ni-* 'put inside', *pane-* 'jump', *ruru-* shake' and *təme-* 'tell' plus a formative *-ŋja*, but the function of this morpheme remains obscure. The extensive truncation of vowels, particularly /a/, across morpheme boundaries in the suffixal sequence of morphemes in the incorporating morphology of the language presented some difficulties in determining the underlying forms of some of the morphemes. The forms presented constitute the current best hypotheses as to what these are, and further work, if any should ever be possible, might result in minor revisions to these. Still I am confident that the description of the language offered here is broadly accurate. Most importantly and as my main motivation, this fascinating language deserves to be described at least as much as possible before it disappears entirely from this world, and that day is not very far off.

Ethnographically, Kopar culture is very similar to that of their Murik relatives; for good descriptions of Murik culture see Lipset (1997) and Schmidt (1922–1923, 1926, 1933). Local oral legends and language distribution tell us a bit about the origins of the Kopar language. Note the following map of the geographical distribution of the Lower Sepik languages (here the Kanda language is labeled Angoram, an alternative name for this language after the district administrative center located in its territory):

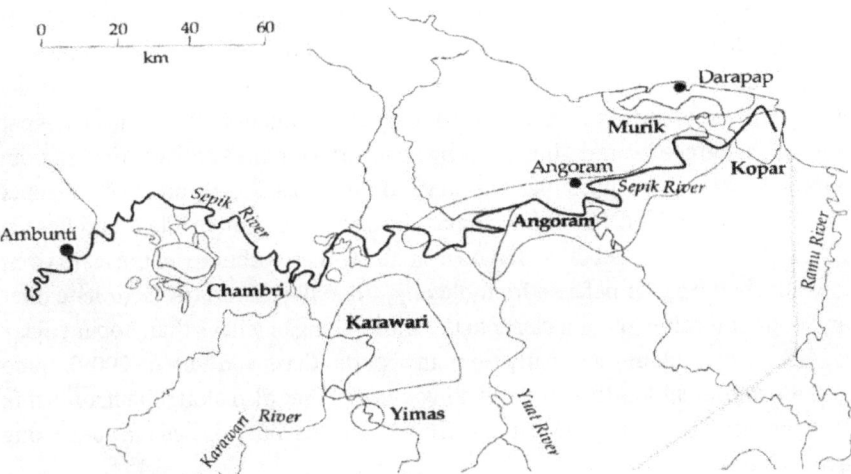

Map 1: Distribution of the Lower Sepik languages.

Note that almost all Lower Sepik languages lie upriver from Kopar (not surprisingly, as Kopar village is situated at the mouth of the Sepik River). The sole exception, Murik, with which it forms a subgroup (see the comparative wordlist of the two languages in Appendix 1), is located in the mangrove swamps to its west; many shared innovations between the two languages demonstrate that they indeed form a subgroup. Given this fact, the ancestral language spoken by Murik and Kopar, Proto-Murik-Kopar, would have constituted a single speech community before their split into two. Legends of the Yimas, located far upriver as Map 1 indicates, clearly narrate a breakup of an ancient speech community including them and the Murik and a consequent migration downriver by the Murik-Kopar ancestors from the homeland of this community which was situated in the high ground upriver from the current location of Yimas village (in confirmation, Chambri legends also tell of a similar migration, this time upriver, from the same area). This clearly indicates an upriver origin of the Kopar language. Legends of the Watam, a village less than ten kilometers to the east of Kopar village right on the coast, add another piece to the puzzle. The Watam language is a Lower Ramu language and its closest relatives like Kaian and Mbore lie still further to its east. Watam legends relate that their ancestors and those of the inhabitants of Kopar village were originally a single community which moved into their current territory from the east. Through barter they arranged to purchase this territory from the village of Singrin, a Kopar speaking village. The ancestors of Watam villagers and Kopar villagers then quarreled, and the ancestral Kopar villagers moved to the current site of Kopar village and adopted the language spoken in Singrin, while ancestral Watam villagers preserved their Lower Ramu language. Singrin seems likely to be the ultimate source village of the three Kopar speaking villages, though the origin of Wongan village is unclear. Overall, dialect data suggest that Wongan and Kopar are closer, and most likely Wongan hived off from a common earlier community and established itself deeper into the mangrove swamps, but this is by no means certain. A split of Wongan from Singrin village is also plausible. Singrin village now sits directly on the main Sepik River on its right bank, but at first contact by German explorers (Claas and Roscoe 2009), it was not located there, but on a small southern tributary of the main river, most likely for easier defense from ongoing intervillage warfare. Both its earlier and current location place it closer to the site of Wongan village than Kopar village is. Also, according to these early German reports (Class and Roscoe 2009), there was an additional fourth Kopar speaking village Potar also along the main Sepik River between present day Kopar and Singrin villages, but this has now been long abandoned.

In terms of its typological profile Kopar is a strongly agglutinative, moderately polysynthetic language, comparable to its sister Yimas (see Foley 2017c). If we adopt the diagnostics for polysynthesis proposed in Foley (2017b), Kopar does qualify as a polysynthetic language, but perhaps not to the degree of some other well known exemplars such as Iroquoian or Caddoan languages. The three diagnostic characteristics of polysynthesis suggested in Foley (2017b) are polypersonalism, head marking and incorporation or dependent-head synthesis (Mattissen 2003). Polypersonalism is the expression of core arguments by agreeing bound pronominals on their governing verb and typically goes hand in hand with head marking. On this, Kopar scores moderately. While basic transitive verbs usually only have agreement for one core argument, so agreement doesn't really qualify as 'poly', they do agree normally through multiple exponence, by spreading out agreement over prefixes and suffixes, so that perhaps rates as 'poly'. Further there is a set of dative pronominal agreement suffixes that commonly occur on verbs, so such verbs do exhibit fuller polypersonalism, with potential agreement for two participants. Besides verbal agreement affixes, Kopar has other characteristics of head marking languages like lack of nominal case marking (Nichols 1986, 2017) and an array of affixes adjusting a verb's arguments, both detransitivizing and transitivizing. Incorporation is also found in Kopar, with rare noun incorporation but much more extensive incorporation of temporals, adverbials and directionals, as well as exuberant verb incorporation. Overall Kopar can be classified as a moderately polysynthetic language.

The phonology of Kopar is fairly typical of its area in the Lower Sepik-Ramu region and shares a number of phonological properties with its distantly related neighbor Watam. With respect to morphology, nouns are simple and verbs are complex, with multiple categories potentially expressed, but at a minimum usually must include tense and person-number of one core argument, subject or object depending on their relative ranking in person on the Animacy Hierarchy (Silverstein 1976; Dixon 1979); hence Kopar like all Lower Sepik family languages belongs to the class of direct-inverse languages. Kopar is more regularly subject-object-verb in clausal word order than Yimas and some other Papuan languages, though permutations do occasionally occur in texts, and postpositional phrases and locatives can follow the verb. This comparative strictness may be due to gaps in the pronominal agreement system for the verb. Clause linkage is mainly done by coordination and subordination, with non-finite constructions for the latter type. Coordination can be simple juxtaposition of full clauses or the more typical Papuan clause chaining, with specialized dependent forms of verbs.

Chapter 2
Phonology

2.1 Phonemes and allophones

Kopar has a somewhat typical phonemic inventory for the languages of the area. The consonants are set out in Figure 2:

		bilabial	apical	lamino-palatal	velar
	voiceless	*p*	*t*		*k*
stops	voiced	(*b*	*d*	*j*	*g*)
	prenasalized	*mb*	*nd*	*ɲj*	*ŋg*
nasals		*m*	*n*		*ŋ*
fricatives			*s*		
rhotics			*r*		
approximants		*w*		*y*	

Figure 2: Kopar consonantal phonemes.

Note that there is no voiceless lamino-palatal stop; as is common in Sepik area languages the /s/ phoneme replaces that. Evidence that supports this claim is that a word initial /s/ following a word final nasal is often realized as the lamino-palatal stop: *indan sur* house inside is often pronounced as [Indaɲjur] and *asandim sur* underneath.of.house inside as [asandimjur]. The /s/ phoneme is commonly realized as an affricate [ts] for older speakers, especially in intervocalic position. Note also that there is no independent lamino-palatal nasal /ɲ/, as this sound only occurs as part of the voiced prenasalized stop /ɲj/. The plain voiced lamino-palatal stop is rare, and particularly so in word initial position; one of the few examples in the data is a borrowed Austronesian word, *jim* 'cloud'. The prenasalized voiced lamino-palatal stop /ɲj/ is more common. These lamino-palatal stops have alternative phonetic realizations as affricates.

The voiceless stops are typically lightly aspirated and more so in initial position, but never as heavily as in English. Impressionistically, they can be quite fortis in their articulation, sounding rather like the voiceless stops of Philippine languages. The plain voiced stops are fully voiced, but they are mostly found in intervocalic position and are marginal phonemes generally derived from the prenasalized voiced stops, for reasons to be explained below. The prenasalized voiced stops are also usually fully voiced, but the stop component of the prenasalized stops can be optionally devoiced in word final position. The prenasalized stops also occasionally weaken their nasal onsets in word initial position. The apical stops have a dental rather than alveolar place of articulation. The velar stops retract toward a uvular articulatory position before the low central-back vowel /a/. Interestingly, the velar voiceless stop does not show a tendency to lenite to [x], as found in many other Papuan, including Lower Sepik, languages; this is probably a result of the fortis articulation of the voiceless stops. The rhotic /r/ is realized as a retroflex flap [ɽ], except following a stop in which case it is tapped [ɾ]. When the coda of a monosyllabic postposition like *sur* 'in' and *tar* 'from', the rhotic is commonly dropped. The phonetics of the remaining phonemes is self explanatory from the choice of the symbols to represent them.

Kopar is a six vowel language, an inventory typical of Lower Sepik languages as well as other languages of the region as exemplified in Figure 3:

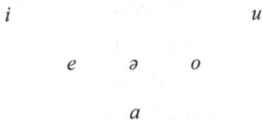

Figure 3: Kopar vowel phonemes.

In addition to these six vowels there are also two common diphthongs /ay/ and /aw/, and a more marginal /oy/ ~ /uy/. The vowels /u/ and /a/ have minimally variable articulations very close to their cardinal articulations: /u/ as in English *sue* [su] and /a/ like the first vowel in *father* [faðə]. /i/ has a variable realization: in open syllables /i/ sounds like the vowel in English *see* [si], but in closed syllables it commonly lowers to its shorter lax variety [I], so sounds closer to the vowel in English *sit*. The mid vowels /e/ and /o/ are less frequent than /i/, /u/ and /a/ and like /i/ have variable realizations depending on syllable types, typically higher and longer [e] and [o] in open syllables, like the vowels in *slay* [sleI] and *slow* [sloʊ] in English, but without the diphthong vocalic offglides, but lower and shorter in closed syllables, [ɛ] and [ɔ], as in English *bet* [bɛt] and *bought* [bɔt]. The vowel /ə/ has the most variable realization, ranging from a lowered high central vowel [ɨ] to a mid central vowel [ə]. It always has the latter realization in open syllables, most notably at the ends of words, but in closed syllables it depends on the place of articulation of the following consonant: higher before apical consonants, lower elsewhere.

The status of the mid central vowel /ə/ has been controversial in Sepik area languages, as it is very commonly used epenthetically to break up disallowed consonant clusters that arise when morphemes are concatenated. Kopar certainly makes use of such an epenthetic /ə/. Consider the behavior of the perfective prefix *t-*. When added to a vowel initial verb, it appears simply as this: *t-Ø-o-n-a* PFV-

3SG-go-LK-PFV 'he/she went'. But when added to one that begins in a consonant, an epenthetic /ə/ must be inserted: *tə-ma-o-n-a* PFV-1SG-go-LK-PFV 'I went'. In some languages like Kalam (Pawley 1966) and Yimas (Foley 1991), the nature of /ə/ epenthesis is so thoroughgoing that all occurrences of /ə/ have been analyzed as epenthetic and the sound removed from the inventory of vowel phonemes, so that vowel-less words are often posited, such as Yimas *tŋkntkn* [təŋgəndəkən] 'heavy'. However, this analysis will not work for Kopar, for which we should distinguish between epenthetic /ə/ and inherent underlying /ə/, and in any case /ə/ is a less frequent sound in Kopar than it is in Yimas. In Kopar there are clear minimal pairs distinguishing words by the presence of /ə/ versus another vowel: *ma* 'I' versus *mi* 'you' versus *mə* 'he, she' (in the Wongan dialect; in the Kopar dialect 'he, she' is *mu); mbə* 'they (DL) versus *mbu* 'they (PL); *məŋgə* 'they (PC) versus *tiŋgi* 'behind'; *nəmbən* 'garamut drum' versus *numbon* 'sago jelly' (*hatwara* in Tok Pisin), *nəmbren* 'pig' versus *nambrin* 'eye', *kənd* 'cassowary' versus *kundə* 'listen!'. A vowel-less analysis of these words is not impossible in Kopar; for example, we could claim that *mbə* 'they (DL) is underlying /mb/, with obligatory epenthetic /ə/ to provide a syllable nucleus, a rule which fails to apply to *mbu* 'they (PL)' since it already has one; but this won't easily account for the contrast between *sond* 'shield' and *məndə* 'feces' or *kənd* 'cassowary' and *kundə* 'listen!' or *rikam* 'bamboo' and *mamə* 'eat!', And there are still other problems with such an analysis, and I will not pursue it here, taking for my descriptive purposes the evidence of the above minimal pairs at face value. One important restriction on /ə/, though, is that unlike the other five vowels, it cannot occur word initially, although /e/ and /o/ are decidedly rare in initial position as well.

2.2 Phonotactics

Kopar words vary from monosyllabic such as *o* 'you (PL) or *nor* 'person' to very polysyllabic, especially in the morphologically complex verbs like *mapipig-abudəkənaya* 'I fished with a hook'. However, roots are usually three syllables or less. The maximal syllable structure of Kopar is CCVC, where only the V slot is obligatory, and words are built up by iterating this syllable structure. Initial segments can be any one of the vowels except /ə/ and one consonant or permissible cluster; the only permissible clusters are those consisting of a voiceless obstuent plus /r/ or /w/. The prenasalized voiced stops freely occur in word initial position, but the plain voiced stops cannot, unless another plain voiced stop follows in the next syllable. Attested word initial vowels are:

/i/	iror	'tree' (the Kopar dialect; the Wongan dialect has *eror*)
/u/	uren	'dog'
/a/	arəm	'water'
/e/	e	'we (PL)' or 'come'
/o/	o	'you (PL) or 'go'

While all these are phonemically vowel initial, as in English they begin with a phonetic glottal stop. Possible consonantal onsets are:

/p/	pətak	'neck'
/t/	tiŋgi	'behind'
/k/	karan	'head'
/s/	surun	'belly'
/b/	bijabətək	'light, not heavy'
/d/	dada	'more'
/j/	jadək	'nothing'
/mb/	mban	'bow'
/nd/	ndesa	'today'
/ɲj/	ɲja	'only'
/ŋgu/	ŋgu	'you (PC)'
/m/	mimiɲ	'tongue'
/n/	naŋgun	'skin'
/ŋ/	ŋaɲjirik	'small'
/r/	ram	'younger brother'
/w/	wakənan	'afternoon'
/y/	yaŋgen	'wallaby'
/pw/	pwap	'middle'
/kw/	kwamb	'Victoria crown pigeon'
/pr/	pre-	'die'
/tr/	traman	'liver'
/sr/	sra-	'cut'
/kr/	krəŋ	'red parrot'

Word final codas are more restricted, and can be a voiceless stop or fricative, a nasal, the rhotic /r/ and a homorganic prenasalized voiced stop, as in these examples:

/p/	karep	'moon'
/t/	imbot	'nose'
/k/	mak	'bad'

/s/	was	'wind' (this word, an Austronesian loan, is the only example)
/mb/	asumb	'mouth'
/nd/	sond	'shield'
/ŋg/	putəŋg	'tree species'
/m/	arəm	'water'
/n/	uren	'dog'
/ŋ/	kinaŋ	'star'
/r/	nor	'person'

The lamino-palatal voiced prenasalized stop cannot serve as a final coda. Nor can the plain voiced stops. As we shall see in the following section on the morphophonemic rules, the latter are not underlying phonemes, but are derived by a mutual dissimilation rule that applies to prenasalized voiced stops in adjacent syllables. Their derived nature blocks them from functioning as word final codas.

Polysyllabic roots are built up by iterating the core syllable structure of CCVC. However, certain constraints apply to this. First of all, a structure of CCVC-CCVC is prohibited by a blanket constraint against any medial consonant cluster consisting of more than two consonants; only CCVCCVC is allowed. Further, medial clusters of two consonants are also heavily restricted. They can either be those licensed in initial position, a voiceless obstruent plus /r/ or /w/, or a nasal plus a stop. Examples of single consonants and consonant sequences in word medial position are:

/p/	napar	'hand'
/t/	itəman	'younger sister'
/k/	kakan	'older same sex sibling'
/s/	sesen	'bird'
/b/	bijabətək	'light, not heavy'
/d/	iduduk	'hot'
/ɟ/	bijabətək	'light, not heavy'
/g/	nigedip	'hourglass drum'
/mb/	imbot	'nose'
/nd/	kundot	'ear'
/ɲɟ/	ŋaɲɟeŋ	'boy'
/ŋg/	niŋgin	'breast'
/m/	maman	'older sister'
/n/	nana	'mother'
/ɲ/	muɲamb	'sugarcane'

/r/	iror	'tree'
/w/	awo	'yes'
/y/	yayan	'father'

/pw/	sapwar	'basket'
/kw/	rukwa-	'cough'
/md/	rakamdan	'night'
/nt/	kintip	'root'
/nj/	akənjim	'rain cloud'
/mbw/	-anumbwa	3PL.DAT
/mbr/	nambrin	'eye'
/ŋgr/	naŋgrin	'spear'

A question arises about the status of the homorganic prenasalized voiced stops: are they clusters as in Yimas (Foley 1991) or unit phonemes? The data argue for the latter for Kopar. Although the phonetic syllable break occurs between the nasal and the stop in medial position, perhaps indicating an analysis as clusters, homorganic prenasalized voiced stops also occur in word initial onset position where there is no syllabic break; ŋgu 2PC is pronounced as a single syllable. Note also that other sequences of nasal plus stop like /nt/ or /md/ cannot occur word initially, only medially; they are clusters, whereas the word initial position of the homorganic prenasalized voiced stops argues that they are unit phonemes. Most importantly, homorganic prenasalized voiced stops occur as codas word finally where otherwise no consonant clusters are permitted at all. Furthermore, if we analyze them as clusters we would need to regard a word such as *nambrin* 'eye' to contain a cluster consisting of three consonants, a pattern otherwise never attested in the language. There is, however, the question of the syllable break in medial position. With medial prenasalized voiced stops, the syllable break is between the nasal and the stop; for instance *imbot* 'nose' is pronounced [ʔIm.bɔt] and *nambrin* 'eye' [nam.brIn]. And this also applies to other medial clusters; for example, *rakamdan* 'night' is [ra.kam.dan], *kintip* 'root' is [kIn.tIp], and *supwar* 'basket' is typically pronounced [sup.war], although [su.pwar] is also possible in this case. Given that the homorganic prenasalized voiced stops are units, not clusters, this syllabification must be regarded as a late level phonetic fact. Consider a monosyllabic word ending in a homorganic prenasalized stop like *kond* 'shield'. This must be syllabified as the following; whatever the phonetic complexities of the homorganic prenasalized stop, it can only occupy a single mora in the bimoraic structure of a monosyllable:

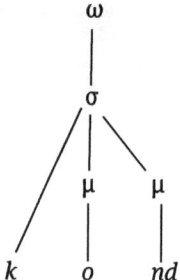

Now consider a disyllabic word like *kundot* 'ear' where the phonetic syllable break is between the nasal and the stop. Given that homorganic prenasalized voiced stops are units, not clusters, they must underlyingly by the universal onset realization rule correspond to the onset of the second syllable:

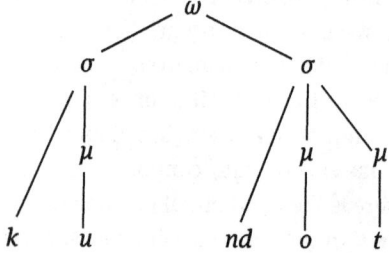

However, given the preference for bimoraic syllables, the syllabification process re-assigns the nasal component of the homorganic prenasalized stop to the previous syllable, as follows:

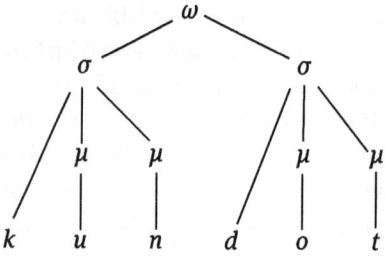

2.3 Stress

Kopar has a distinction between stressed and unstressed syllables. Phonetically stressed syllables are longer and higher in pitch than unstressed ones, but to a noticeably lesser degree than in English. Nor are unstressed syllables reduced as in English. Stress is not predictable in Kopar, although I have identified no minimal pairs, i.e. words distinguished solely by the placement of stress. In disyllabic words stress is usually final: *a'sumb* 'mouth', *i'ror* 'tree', *ki'naŋ* 'star', *a'kən* 'sun', *ka'rep* 'moon', *mi'miŋ* 'tongue', *pə'kəp* 'flying fox, *ra'ri* 'yesterday/today'. But this is not always so: *'yayan* 'father', *'nana* 'mother', *'yaŋgen* 'wallaby', *'naɲjen* 'boy', *'ndesa* 'today' *'tambək* 'five', *'traman* 'liver'. For trisyllabic words, again, final stress is most common:, *itə'man* 'younger sister', *asi'rap* 'tooth', *naŋgi'tap* 'finger', *nəmbi'raŋ* 'leaf', *ambi'sen* 'girl', *ara'tək* 'good'. But a few have penultimate stress, *nə'mandək* 'woman', *ra'rindək* 'yesterday', *pa'tendik* 'old', *sa'nandək* 'four' (though all of these could be analyzed as having final stress with a suffix -ndək), and some initial stress, *'nigidip* 'hour glass drum', *'menome* 'who', *'kiriŋgip* 'armlet', *'mbiona* 'one', *'petəndək* 'black'. Words of four syllables or more are rare as underived roots, although very frequent in the morphologically rich verbs. Primary stress in the often long polysyllabic multi-morphemic verbs is penultimate: *ma-mə-ndək-naya* 1SG-eat-NR.PAST-1SG 'I ate' [ˌma.mə.ˌndə.kə.'na.ya], with secondary stresses on alternating preceding syllables, unless the penultimate syllable contains an epenthetic schwa. In that case, the primary stress goes on the antepenultimate syllable: *ku-samayt-ka* TR.IMP-see-2SG [ˌku.sa.'may.tə.ka] 'watch him!'. Stress in the few non-verbal forms attested with four syllables always have final stress, but with a secondary stress on the antepenultimate syllable: *kaˌŋgona'sen* 'shark', *kuˌŋgopa'rik* 'long', *biˌjabə'tək* 'light (in weight)', *saˌrapa'kin* 'heat'.

2.4 Morphophonology

There are a number of morphophonemic rules that operate in Kopar, and because elaborate morphology in Kopar is a feature of verbs, their main domain of operation is across agglutinating affixes of verbs. These rules affect both consonants and vowels. Let me consider the rules that apply to vowels first. Like Yimas (Foley 1991), Kopar has a rule of vowel harmony, resulting in both the raising and fronting of /ə/ in the environment of a front vowel and its raising and backing in the presence of a back vowel. Consider the following two inflected intransitive forms of the verb *mə-* 'eat':

(2.1) (a) *mamǝndǝkǝnaya*
ma-mǝ-ndǝk-naya
1SG-eat-NR.PAST-1SG
'I ate'

(b) *umundukoya*
u-mǝ-ndǝk-oya
3SG-eat-NR.PAST-3SG
'he ate'

The verb root *mǝ-* 'eat' contains an underlying /ǝ/ and is here inflected with the NR.PAST tense suffix *-ndǝk*. Following the tense suffix *-ndǝk* is found the first singular subject agreement suffix *-naya*. An epenthetic /ǝ/ is inserted in (2.1a) to break up impermissible consonant cluster resulting from the accretion of the two suffixes, between the final /k/ of *-ndǝk* and the initial /n/ of the subject agreement suffix *-naya* 1SG. In (2.1b) again there are two /ǝ/s, that of the verb root *mǝ-* 'eat' and that of the tense suffix *-ndǝk*. The subject agreement suffix in this case is *-oya* third person singular. The initial vowel of this suffix is /o/, a rounded back vowel. This vowel triggers the vowel harmony rule, causing the /ǝ/ of the tense suffix *-ndǝk* to raise and round to /u/. This derived /u/ of now *-nduk* is also a back rounded vowel and in turn causes the /ǝ/ of the verb root *mǝ-* 'eat' to raise and back to /u/. The third singular prefix is underlying *u-* and is therefore unaffected. Now consider the form in (2.2):

(2.2) *paŋgǝ imindikiya*
paŋgǝ i-mǝ-ndǝk-iya
1PC 1-eat-NR.PAST-PC
'we (PC) ate'

Here the paucal agreement suffix *-iya* PC begins in a high front vowel. It also triggers the vowel harmony rule, causing the underlying /ǝ/ in both the preceding suffix and the verb root to raise and front to /i/. First the vowel of the tense suffix is fronted to /i/, and the tense suffix now containing a high front vowel causes the /ǝ/ of the verb root *mǝ-* to front to /i/. Again the verbal prefix for a non-singular first person is *i-*, so remains unaffected. Note that these vowel harmony rules converting /ǝ/ into either /i/ or /u/ indicate that /ǝ/ is regarded phonologically as a high vowel in Kopar, albeit one unspecified for frontness.

These data require the postulation of two morphophonological rules, one for /ǝ/ epenthesis and one for vowel harmony:

(2.3) /ə/ Epenthesis
Ø → ə / C_C
where CC is anything other than the permitted consonant clusters of section 2.2

(2.4) Vowel Harmony
ə → [α front] / ___ C [α front]
                 V

This rule converts /ə/ whether epenthetic or underlying into /u/ when the vowel of the following syllable is /u/ or /o/ and into /i/ when it is /i/ or /e/.

The vowel harmony rule sometimes fails to apply, and this more commonly is the case with fronting rather rounding. The rule also often fails to be enacted recursively, only applying once, so that, for example, an acceptable alternative to (2.2) would be (2.5):

(2.5) paŋgə iməndikiya
     paŋgə i-mə-ndək-iya
     1PC   1-eat-NR.PAST-PC
     'we (PC) ate'

where the vowel harmony rule fails to apply to the root's /ə/. Also phonetically, /ə/ is usually realized as a syllabic nasal in a sequence of an obstruent and a prenasalized voiced stop: *samayt-ndək* see-NR.PAST [samaytn̩dək].

The /ə/-epenthesis rule is due to constraints of syllabification. Consider the underlying form of example (2.1a):

CV CV CV C CV CV
| | | | | | | | | | |
ma-mə-ndə k –na ya

Syllabification is straightforward from the beginning of the word until we come to the consonant cluster /kn/. This cannot syllabified as a coda of a putative syllable /ndəkn/ because syllables do not permit more than a single consonant as codas. Nor can it be taken as an onset of a putative syllable /kna/ because this is not an allowed onset consonant cluster in Kopar. But neither /k/ nor /n/ can be left unsyllabified. Hence the /ə/-epentheis rule applies, inserting a vowel facilitating syllabification:

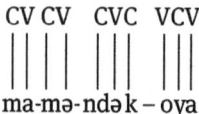

The vowel harmony rule operates by spreading the association of the features [+/-front] to the vowel /ə/, either underlying or epenthetic, in adjacent syllables. Consider example (2.1b). Underlyingly it is:

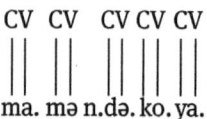

There are no impermissible consonant clusters here so syllabifcation proceeds without /ə/ epenthesis; if there were, /ə/ epenthesis must precede vowel harmony:

CV CV CV CV CV
| || || || ||
ma. mə n.də. ko. ya.

Now, vowel harmony applies by spread the feature specification of [-front] of /o/ to the /ə/ of the preceding syllable converting it to /u/, and now the [-front] feature of this /u/ spreads to the /u/ in the syllable preceding it, also converting it to /u/:

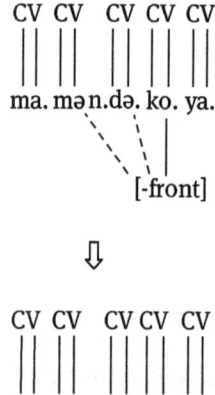

Note that vowel harmony can proceed leftward as above, the more common case, but also rightward. Consider the word for 'belly, stomach' *surun*, which is derived from the postposition *sur* 'inside, in' plus the nominalizing suffix *-n*; hence its underlying form is:

/rn/ as a consonant cluster is not permitted as a coda, so /ə/ epenthesis must apply to allow proper syllabification:

Now vowel harmony applies rightward, spreading the [-front] feature of /u/ to /ə/:

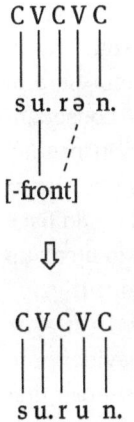

Kopar always deletes underlying /ə/ in the immediate environment of a following suffix beginning in a vowel (no suffix can begin in /ə/). But Kopar commonly prohibits vowel coalescence and truncation, except in the case of identical vowels coming together, and not even always there (see rule 2.9). Commonly, a previous [-low] vowel desyllabifies to its corresponding approximant, but the conditions for this are still rather murky, as it sometimes fails to apply when it might be expected. For example *u-si-ar-ana-k* 3SG-do-PROG-3SG.DAT-FR.PAST 'he was feeling it' is pronounced [usiaranak] not *[usyaranak] as might be expected (compare 2.6b). Here are some examples of desyllabification:

(2.6) (a) mənda   iprarangəbakwe?
         mənda   i-pra-ar-aŋg-mbako-e
         feces   2-excrete-PROG-PRES-2DL-Q
         'are you (DL) defecating?'

   (b) petəndək   təsya
       petəndək   t-Ø-si-a
       black   PFV-3SG-become-PFV
       'it has turned black'

   (c) uyaraŋoya
       u-e-ar-aŋg-oya
       3SG-come-PROG-PRES-3SG
       'he is coming'

(2.7) Desyllabification
           V → [-syllabic] / ___ V
           [-low]

Desyllabification fails to occur if an initial consonant cluster that would result from it is impermissible. For instance, *ŋgi-o-k* 3PL-go-FR.PAST cannot be pronounced as a single syllable [ŋgyok], as [ŋgy] is not a permitted initial consonant sequence. However, in contexts like this, a phonetic approximant is often inserted between the two vowels, hence [ŋgiyok]. This could be analyzed as a low level phonetic phenomenon of transition between a high front vowel and a mid back vowel, or disyllabification as in rule (2.7) followed by /ə/ epenthesis to break up the impermissible initial consonant cluster followed by vowel harmony triggred by the [+front] approxomant, i. e.: *ŋgi-o-k* > ŋgyok (desyllabification) > ŋgəyok (/ə/ epenthesis) > ŋgiyok (vowel harmony). There is no convincing evidence to choose between these alternatives given the available data, but the simpler first hypothesis seems overall preferable.

Desyllabification in some ways is the inverse of /ə/ epenethesis. Here vowels lose their status as syllable nuclei, thereby deleting a syllable from the word. Consider the derivation of (2.6c)

V V V C V C V C V
| | | | | | | | |
u e a r a ŋgo y a

This sequence of three vowels is impermissible and this is truncated by desyllabifying the medial [+front] vowel /e/ and turning it into its corresponding [+front] approximant /y/ and hence consonantal onset by rule (2.7):

V C V C V C V CV
u y a r a ŋg o ya

So now syllabification applies straightforwardly:

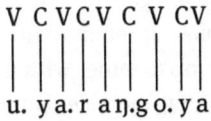

V C V C V C V CV
u. ya. r aŋ.g o. ya

There is one glaring exception to desyllabification and that concerns a few verbal suffixes that begin in /a/ like the singular perfective -a. When these follow a root or another suffix that ends in a [-high] vowel, an /n/ is inserted between the two vowels; compare:

(2.8) (a) petəndək   təsya
        petəndək   t-Ø-si-a
        black   PFV-3SG-become-PFV
        'it has turned black'

   (b) tona
       t-Ø-o-a
       PFV-3SG-go-PFV
       'he has gone'

   (c) təma
       t-Ø-mə-a
       PFV-3SG-eat-PFV
       'he has eaten'

   (d) təprena
       t-Ø-pre-a
       PFV-3SG-die-PFV
       'he has died'

(2.9) /n/ Insertion
∅ → n / V__a   (for certain verbal suffixes)
[-high]

Note, as (2.8c) illustrates, that /ə/ counts as a [+high] vowel here: no /n/ insertion. That is in keeping with its overall phonological properties in the language, i.e. when it is fronted it is realized as /i/ and when rounded as /u/, both [+high] vowels. Instead it is deleted here via the general rule of /ə/ deletion before a suffix beginning in a vowel.

The most striking and crosslinguistically unusual morphophonological rule in Kopar is the one that affects prenasalized voiced stops. There is a general prohibition in the language against these being in adjoining syllables. When morphological formations bring an affix with a prenasalized voiced stop together with a root or affix containing another prenasalized stop in the adjoining syllable, both it and the one in the adjoining syllable are denasalized into the corresponding plain voiced stops, a mutual dissimilation rule in a sense, but one that results in assimilating each to the other. This process is widespread in Kopar verbal paradigms. Consider the following examples:

(2.10) (a) kari**mbi**ya
kar-mbi-oya
walk-IM.FUT-1SG
'I will walk'

(b) kar ŋga ri**ndu**ku
kar ŋga ri-onduku
walk DAT DES-2PC
'you (PC) want to walk'

(c) kari**bidu**ku
kar-mbi-onduku
walk-IM.FUT-2PC
'you (PC) will walk'

Example (2.10a) illustrates the basic form of the immediate future tense suffix -*mbi*, while (2.10b) does the same for second person paucal subject agreement suffix for the unrealized tenses -*nduku*. Example (2.10c) is the immediate future tense for a second person paucal subject. Note that both the immediate future tense -*mbi* and the subject suffix -*nduku* are denasalized, resulting in -*bi* and -*duku* respectively. Now consider these examples:

(2.11) (a) ŋgikarəndəkəŋgaya
ŋgi-kar-ndək-ŋgaya
3PL-walk-NR.PAST-3PL
'they walked'

(b) ukararaŋgoya
u-kar-ar-aŋg-oya
3SG-walk-PROG-PRES-3SG
'he is walking'

(c) ŋgikaragəgaya
ŋgi-kar-ar-aŋg-ŋgaya
3PL-walk-PROG-PRES-3PL
'they are walking'

Example (2.11a) exemplifies the basic form of the third plural subject agreement suffix for intransitive verbs -ŋgaya, and example (2.11b) the present tense suffix -aŋg. Example (2.11c) is the form for the present tense for intransitive verbs with third person plural subjects. Again, both the present tense suffix -aŋg and the third person plural suffix -ŋgaya are denasalized to -ag and -gaya respectively.

We can formalize this mutual denasalization rule as follows:

(2.12) [+stop]   V   [+stop]   →   [-nasal]   [-nasal]
        [+voice]   [+voice]
        [+nasal]   [+nasal]

i.e. any sequence of two prenasalized voiced stops in adjacent syllables lose their prenasalization. Speakers fail to apply this rule in every case where its conditions are met, though its application in the near immediate tense, for example, is unexceptional. Sometimes only one of the voiced prenasalized stops, typically the second, is denasalized (see example 2.6a), yet its application is still widespread. This rule accounts for one very unusual typological fact of the language: with rare exceptions, plain voiced stops are only encountered in Kopar when there are two or more of them in adjacent syllables, for example: *bijabətək* 'light, not heavy', *iduduk* 'hot', *nigedip* 'hourglass drum', *jadək* 'nothing', *pipigabu-* 'fish with a hook'. This phonotactic constraint suggests that there may not be any underlying voiced stops in Kopar at all, but that they are derived from the denasalization rule of (2.12). One piece of evidence for this comes from the inflection of *pipigabu-* 'fish with a hook'. Example (2.13) shows that verb root inflected for first singular subject in NR.PAST tense:

(2.13) *mapipigabudəkənaya*
ma-pipiŋgambu-ndək-naya
1SG-fish.with.hook-NR.PAST-1SG
'I fished with a hook'

Note that here the NR.PAST tense suffix -*ndək* surfaces in the denasalized form -*dək*, but for that to be the case the onset consonant in the preceding syllable needs also to be underlyingly a prenasalized voiced stop and in turn the same must hold for the consonant of the syllable proceeding that. Therefore, the root *pipigabu* must underlyingly be pipiŋgambu-. The form of the verb root is the Wongan dialect is exactly that, subject to mutual denasalization, *pipigabu-*, but speakers of the Kopar dialect typically avoid applying denasalization here by devoicing the /ŋg/ and produce a variant *pipikambu-* instead. It should also be pointed out that the mutual denasalization rule of (2.12) is also found in the neighboring, but only quite distantly related, Lower Ramu language Watam, where it is unexceptional and even more pervasive. Whether this is due to a retention from the earlier shared history of speakers of these two languages or an innovation within this small linguistic area is unknown.

Finally, there is an optional, but rather common rule that is somewhat the opposite of denasalization. This is a rule that allows a final nasal in a verb or noun root to fortify into a homorganic prenasalized voiced stop, so that roots can alternate between a nasal or a prenasalized voice stop finally, for example *nəman ~ nəmand* 'woman', *kinim- ~ kinimb-* 'to fasten', *Ombaŋ ~ Ombaŋg* 'PN'. Given that there is a phonemic contrast with roots that already end in a homorganic prenasalized voiced stop, this needs to stated formally as a rule that applies to roots ending in a nasal:

(2.14)   C   →   C   / __ #
       [+nasal]   [+nasal]   +
                       [+stop]

2.5 A note on presentation of examples

Due to the complex morphophonemics of the verbs, the orthography that I will employ in the remainder of this grammar sketch for the presentation of examples will typically follow the pattern established here. It will consist of four lines, the first, in italics, a standard phonemic transcription, including any epenthetic /ə/'s, given its apparent phonemic status in Kopar; the second, the underlying representation before the general morphophonemic rules of section 2.4 or those specific to verbal affixes described in Chapter 5 apply, although any epenthetic /n/ added by rule (2.9) will be written to aid legibility; the third, a morpheme by morpheme gloss, with epenthetic /n/ glossed as LK; and the fourth, a translation into English.

Chapter 3
Word classes

Kopar has only two major classes of words, nouns and verbs; all other classes are small and closed. Like Papuan languages generally, there is a clear morphosyntactic contrast between nouns and verbs. The inventory of the small closed classes is:

adjectives	quantifiers
pronouns	deictics
postpostions	temporals
conjunctions	interjections

Like other Lower Sepik languages such as Yimas (Foley 1991) and Murik (Schmidt 1953), the set of uncontroversial adjectives in Kopar is very small; most words denoting properties share a number of the morphosyntactic features of verbs.

I will use a mixture of semantic and formal criteria to define each class. Consider nouns. Semantic criteria for the class of nouns include that they prototypically denote objects. But some do not, expressing instead states or properties such as *iŋarapin* 'shame, ashamed' or *sarapakin* 'heat'. Semantic criteria for verbs include that they prototypically denote events. Again, some do not; they denote states, such as *si-* 'feel, want', *ndat-* 'know'. Hence formal properties are needed to further precisely define the distinction between nouns and verbs. Formally, nouns function as syntactic arguments of verbs, and as such verbs can flag the number of their nominal arguments when functioning in subject or object grammatical function. Verbs as predicates and governors of their arguments flag the person and number of their subject or object grammatical functions and further require tense marking, situating the time of the utterance in the temporal continuum. Example (3.1) illustrates these features for nouns and verbs:

(3.1) *nəmandəpak ŋgiok*
 nəmandəpak ŋgi-o-k
 woman.PL 3PL-go-FR.PAST
 'the women went'

The subject argument of this sentence is *nəmandəpak* 'women', and, as expected, the verb agrees with it as *ŋgi-*, the third person plural subject pronominal agreement prefix for intransitive verbs like *o-* 'go'. And the verb, as required, is additionally inflected for tense, here the far past tense suffix *-k*.

3.1 Nouns

Unlike in its sister languages Yimas and Kanda, nouns in Kopar are morphologically simple. There is no gender or noun class marking, and number marking is facultative and rare for all nouns except those with human referents. Nor, as a head marking language, do the arguments of a verb in Kopar bear case. Hence, there are no strictly morphological inflectional properties for nouns that can be used to define this class. There are, however, a few syntactic properties. The first was already mentioned above: nouns function as the arguments of verbs; or, more precisely, noun phrases functioning as the arguments of a verb can be headed solely by a noun and nothing else. In fact, expanded noun phrases are relatively rare in Kopar discourse, as the majority of noun phrases consist of just a bare noun. Examples in (3.2) illustrate:

(3.2) (a) nor   karobida
        nor   kar-o-mbi-onda
        man   walk-go-IM.FUT-3SG
        'the man will walk around'

    (b) nor   nəmbren   mbunambrataŋgoya
        nor   nəmbren   mbu-nambrat-aŋg-oya
        man   pig   3.ERG-spear-PRES-3SG
        'the man spears the pig'

    (c) nor   puruŋ   nəmandək   ŋga   mbutukamaŋgoya
        nor   puruŋ   nəmandək   ŋga   mbu-t-kam-aŋg-oya
        man   betelnut woman   DAT   3.ERG-CAUS-arrive-PRES-3SG
        'the man gives betelnut to the woman'

The examples in (3.2) illustrate one argument (3.2a), two argument (3.2b) and three argument (3.2.c) verbs with their subcategorized arguments. Note that there is no true case marking; subject and object arguments are always unmarked, while the recipient of a three argument verb here bears the dative postposition ŋga. Functioning bare like this as arguments with no inflection or derivation is a unique defining feature of nouns; no other part of speech can do this. When verbs function as nouns they are nominalized via the nominal suffix *-n: tra-* 'to dance' > *tra-n* 'a dance'.

Related to this is that nouns or their extension in noun phrases can function as the complement of postpositions. Example (3.2c) illustrated this with the dative postposition, and here are a couple of other examples:

(3.3) (a) okoya   inda   sur
           o-k-oya   inda   sur
           go-FR.PAST-3SG   house   inside
           'he went inside the house'

    (b) nor   iror   kaŋgarap   təndasana
         nor   iror   kaŋgarap   t-Ø-ndasa-n-a
         man   tree   above   PFV-3SG-sit-LK-PFV
         'the man has sat on top of the tree/wood'

3.2 Verbs

Verbs are far and away the most morphologically complex class of words in Kopar and they are highly so. This is completely in accord with the head marking typology of the language. All other word classes are morphologically simple. Chapter 5 will be entirely devoted to the morphology of verbs, so I will only discuss some basic defining features for this word class here.

Finite verbs require two morphological categories to be specified (non-finite verbs are nominalized and therefore morphologically simple like nouns). These categories are tense and person and number of their subject or object arguments. No other part of speech is marked for these inflectional categories. Both of them are represented primarily by suffixes, though there are prefixes as well. Kopar has a complex tense system, consisting of a present tense and multiple past and future tenses, illustrated in the following examples:

(3.4) (a) mibida
           mə-mbi-onda
           eat-IM.FUT-3SG
           'he should eat soon'

    (b) umondukonda
         u-mə-ondək-onda
         3SG-eat-FUT-3SG
         'he will eat (later)'

    (c) umaŋgoya
         u-mə-aŋg-oya
         3SG-eat-PRES-3SG
         'he eats now'

(d) *umundukoya*
u-mə-ndək-oya
3SG-eat-NR.PAST-3SG
'he ate yesterday'

(e) *umukoya*
u-mə-k-oya
3SG-eat-FR.PAST-3SG
'he ate before yesterday'

(f) *təma*
t-Ø-mə-a
PFV-3SG-eat-PFV
'he has eaten'

The verbal indication of the person and number of a subject or object argument is extremely complex in Kopar, and much of Chapter 5 will be devoted to this topic. It is determined by the tense-aspect of the verb, the verb's transitivity and the relative ranking of the subject argument with respect to the object argument on the Animacy Hierarchy (Silverstein 1976; Dixon 1979). Note, for example, in (3.4), the contrast for the third singular subject marker in the immediate future and future tenses *-onda* (becoming *-da* after vowel deletion and denasalization in the immediate future) versus *-oya* for other tenses versus Ø for the perfective. First person singular is similar:

(3.5) (a) *mimbiya*
mə-mbi-oya
eat-IM.FUT-1SG
'I should eat soon'

(b) *mamaŋgaya*
ma-mə-aŋg-aya
1SG-eat-PRES-1SG
'I eat'

(c) *təmama*
t-ma-mə-a
PFV-1SG-eat-PFV
'I've eaten'

(d) *maməndəkənaya*
ma-mə-ndək-naya
1SG-eat-NR.PAST-1SG
'I ate'

Transitivity of the verb also determines the inflectional pattern for agreement. Intransitive and transitive verbs look very different:

(3.6) (a) *nəmandəpak ŋgikarəndəkəŋgaya*
nəmandəpak ŋgi-kar-ndək-ŋgaya
woman.PL 3PL-walk-NR.PAST-3PL
'the women walked'

(b) *nəmandəpak nor mbusamaytundukondu*
nəmandəpak nor mbu-samayt-ndək-ondu
woman.PL man 3.ERG-see-NR.PAST-3PL
'the women saw the man'

Note that the subject agreement suffixes for the third plural subjects are very different for the intransitive verb *kar-* 'walk' and the transitive verb *samayt-* 'see': *ŋgi-...-ŋgaya* versus *mbu-...-ondu*. This pattern is general, not restricted to third person; here are examples with second singular:

(3.7) (a) *mi uri ŋga iraŋgaya*
mi uri ŋga i-ru-aŋg-aya
2SG crocodile DAT 2-shoot-PRES-2SG
'you (SG) shoot at a crocodile'

(b) *mi uri inambrataŋgona*
mi uri i-nambrat-aŋg-ona
2SG crocodile 2-spear-PRES-2SG
'you (SG) spear a crocodile'

The verb *ru-* 'shoot at, hit' in (3.7a) is intransitive, as indicated by the dative postposition *ŋga* marking the target of the shooting, *uri* 'crocodile'. The verb bears intransitive subject marking for second person *-naya*. In (3.7b) the verb *nambrat-* is fully transitive, and as expected, both the subject and object arguments are bare. The verb now bears the subject marker for second singular for transitive verbs *-ona*. Fundamentally, there is an ergative system underlying these patterns, but it is quite complicated, and its full elucidation will need to wait for Chapter 5.

Finally, the subject's relative ranking in person along the Animacy Hierarchy against that of the object can determine its realization. Consider the case when a second person subject acts on a third person object (3.8a) versus when it acts on a first person object (3.8b):

(3.8) (a) mi nor isamaytəndukona
mi nor i-samayt-ndək-ona
2SG man 2-see-NR.PAST-2SG
'you (SG) saw the man'

(b) mi ma ŋgasamaytəndəkənaya
mi ma ŋga-samayt-ndək-naya
2SG 1SG INV-see-NR.PAST-1SG
'you (SG) saw me'

What is crucially at work here is a direct-inverse system as in Algonkian languages like Plains Cree (Dahlstrom 1991; Wolfart 1973; Wolvengrey 2011; see also Jacques and Antonov 2014) overlaying an ergative agreement schema, a pattern common to Lower Sepik languages generally: hence the higher ranked subject agrees on the verb in (3.8a), but the higher ranked object does so in (3.8b) and the subject fails to agree. The details are too complex to go into here and will be treated in depth in Chapter 5. The point to be noted is that the expression of agreement for core arguments depends on relative ranking of person.

Another unique defining property of verbs beside the obligatory marking of tense and subject person-number is the indication of mood, such as imperative. Only verbs can be inflected for imperative mood, and again intransitive verbs and transitive verbs are inflected differently:

(3.9) (a) intransitive
marəməka
ma-rəmə-ka
ITR.IMP-stand-2SG
'stand up!'

(b) transitive
kusamaytəka
ku-samayt-ka
TR.IMP-see-2SG
'see it!'

The prefix *ma-* is used for imperatives of intransitive verbs and *ku-* for transitive verbs.

Finally, only verbs can be negated. There are two negative verbs, a general *kay-* and a prohibitive or negative imperative *nda-*. Although overtly tenseless, both are inflected with the person-number subject agreement suffixes of the paradigm for the unrealized tenses. Note that the subject agreement suffixes in these cases are floated away from the main verb and realized on the negator:

(3.10) (a) mu   rari   ndonduk   kaynda
           mu   rari   ndə-ondək   kay-onda
           3SG   1.day.removed   hear-FUT   NEG-3SG
           'he won't listen tomorrow'

       (b) mi   arəm   kiri   ndana
           mi   arəm   ki-ri   nda-ona
           2SG   water   wash-DOWN   PROHIB-2SG
           'don't bathe!'

3.3 Adjectives

The question of the adjective class in Kopar cannot be completely settled on the basis of the limited data available. There is at least one true adjective, but whether there are more depends on how we interpret conflicting data. The situation is somewhat similar to that of Yimas (Foley 1991:93–101). Words denoting properties can be divided into three classes:

Adjective:
    kapu ~ kapa 'big'

Adjectival verbs:
    aratək 'good'   *mak* 'bad'
    kaymbak 'light, white'   *petəndək* 'dark, black'
    ikuduk 'red'   *ŋaɲjirik* 'small'
    kuŋgoparik 'long'   *katarik* 'short'
    patəndək 'heavy'   *bijabətək* 'light in weight'
    ududuk 'hot'   *pəsayk* 'cold'
    patendək 'old'   *nəŋgəmək* 'new'

Nouns:

iŋarapin 'shame'	*sarapikin* 'cold'
nime 'hunger'	*kanda* 'sickness'
ŋgeroŋ 'happiness'	*imbotma* 'jealousy'

Property denoting nouns denote experiences and therefore occur in experiential clauses (see section 6.1.2.4). These are impersonal, so verbs do not agree with property nouns; such verbs have third person singular subjects:

(3.11) ma iŋarapin təsya
ma iŋarapin t-Ø-si-a
1SG shame PFV-3SG-do-PFV
'I feel ashamed'

The lexemes for 'hungry' *nime* and 'sickness' *kanda* are complicated in that they do not seem to occur on their own as simple monomorphemic lexemes when nominal arguments, but must be buttressed with other head nouns in compounds, *surun* 'belly' for *nime* 'hunger' and *mora(n)* 'thing' for *kanda* 'sickness'.

(3.12) (a) ma nime surun təsya
ma nime surun t-Ø-si-a
1SG hunger belly PFV-3SG-do-PFV
'I feel hungry'

(b) nana yo mora kanda sisirik
nana yo mora kanda si-siri-k
mama DEF thing sick do-MORNING-FR.PAST
'mama felt sick in the morning'

In neither of these cases do these words denoting properties have the characteristics of an adjective. An adjective follows its head noun in Kopar, but *nime* 'hunger' precedes the noun *surun* 'belly', and, while *kanda* follows its head *mora* 'thing', it shows an alternation with or without a final /n/, which is a diagnostic of nouns (see section 4.1). In fact, these examples look a good deal like a noun-noun compound such as *uren-ŋaɲje(n)* dog-boy 'puppy'.

The most striking contrast between the sole uncontroversial adjective *kapu* 'big' and the adjectival verbs is the presence of a suffix *-k* in all of the latter. This suffix is cognate with one of the same form in Yimas and has the same distribution (Foley 1991:237). In Kopar it is a far past tense suffix, meaning from the day before yesterday to any time beyond that, and is commonly used in narrating traditional stories and legends, events found in a time out of living memory (3.13a) to indicate that the time of the utterance is not bound in the normal temporal continuum. In this usage I gloss it as far past, FR.PAST. It is also used for events which are hypothetical and hence not in the temporal continuum (3.13b), and for properties, which are states not bound in time, i.e. the adjectival verbs; in these usages I will gloss it as irrealis IRR:

(3.13) (a) nəmbən   budukondu
nəmbən   mbu-ndə-k-ondu
garamut.drum   3.ERG-hear-FR.PAST-3PL
'they (PL) heard the garamut drum'

(b) sokay   mə-kə-mb   o   mak   simbina
sokay   mə-k-mb   o   mak   si-mbi-ona
tobacco   eat-IRR-OBL   bad   become-IM.FUT-2SG
'if you (SG) smoke tobacco, you (SG) will get ill (literally 'become bad')

(c) naŋgun   ma-na   kaymbakoya
naŋgun   ma-na   kaymba-k-oya
skin   1SG-POSS   white-IRR-3SG
'my skin is white'

Note that in this analysis the morphological structure of the property denoting word 'white' in (3.13c) is identical to that of the event denoting verbs 'hear' and 'eat': a tense marker -k IRR and a subject agreement suffix. Note this is not possible for the one uncontroversial adjective *kapu* 'big':

(3.14) (a) indan   kapu   yo
indan   kapu   yo
house   big   DEF
'the big house'

(b) *indan   ma-na   kapukoya
indan   ma-na   kapu-k-oya
house   1SG-POSS   big-IRR-3SG
'my house is big'

Another difference is in the intensifiers. *Kapu* 'big' takes *suman* 'very' following it, while the adjectival verbs occur with *nuŋgo* 'very', preceding them in the way manner adverbs modifying verbs do (3.15c):

(3.15) (a) tamənd yo kapu suman
   tamənd yo kapu suman
   fish DEF big very
   'the fish is very big'

(b) nəmbren yo nuŋgo ŋaɲjirikoya
   nəmbren yo nuŋgo ŋaɲjiri-k-oya
   pig DEF very small-IRR-3SG
   'the pig is very small'

(c) nor, aratək marəməka
   nor, aratək ma-rəmə-ka
   man good ITR.IMP-stand-2SG
   'hey man, stand up straight!'

Note that *nuŋgo* 'very' occurs in the same position immediately before its property denoting word as the manner adverbial *aratək* 'good, well' does before its verb, while *suman* 'very' is placed after *kapu* 'big'. This argues that *ŋaɲjiri-* 'small' and *rəmə-* 'stand' belong to the same word class, verbs, while *kapu* belongs to a different one, adjectives.

However, if adjectival verbs are verbs, they are defective verbs. In (3.15c) the adjective verb *aratək* 'good' is used adverbially to modify the main verb. But under no circumstances can it occur here with subject agreement suffixes, as true verbs do. This suggests that it is not a full verb. Nor does the *-k* seem to function any longer as a tense suffix; the main verb is in imperative mood. In fact, such a sequence of two full verbs is ungrammatical:

(3.16) *nor, aynde sirik marəməka
   nor, aynde siri-k ma-rəmə-ka
   man here descend-IRR ITR.IMP-stand-2SG
   'hey man, come down and stand here!'

These data indicate that whatever the diachronic origin of *-k* on adjectival verbs, it no longer is a productive suffix, but has been re-analyzed as an inherent part of the adjectival verb. Therefore, it would follow that adjectival verbs do not now bear tense suffixes, as all true verbs must do, and so should be regarded as distinct from them. Yet they are distinct from *kapu* 'big' as well. They still bear a number of properties of verbs, so I regard adjectival verbs as a type of defective verb, distinguishing them from true verbs, on the one hand, and from the one true unequivocal adjective *kapu* 'big', on the other.

There are other properties that distinguish adjectival verbs from true verbs. Adjectival verbs never have agreement when functioning attributively (3.17a), and even when functioning predicatively, they sometimes lack it (3.17b), though not if their subject is first or second person (3.17c). Lack of suffixal subject agreement on verbs, although permissible, is rare out of context (it is usual though on dependent verbs in clause chaining constructions), and requires an anaphoric context to license it (3.17d); no such contextual reading is needed for agreement-less adjectival verbs when their subjects are third person:

(3.17) (a) nəmbren petəndək-(*oya) yo
nəmbren petəndək-(*oya) yo
pig black-3SG DEF
'the black pig'

(b) nəmbren yo petəndək-(oya)
nəmbren yo petəndək-(oya)
pig DEF black-3SG
'the pig is black'

(c) ma nor kuŋgoparikə-*(naya)
ma nor kuŋgoparik-*(naya)
1SG man long-1SG
'I'm a tall man'

(d) nəmbren yo uprenaraŋ-?(oya)
nəmbren yo u-pre-n-ar-aŋ-?(oya)
pig DEF 3SG-die-LK-PROG-PRES-3SG
'the pig is dying'

And there are still other constructions which distinguish the adjective and the adjectival verbs from true verbs, for instance specialized inchoative and causative constructions, which are available for both classes, but not for true verbs. The adjective and adjectival verb denote a resultant state in these constructions which require the verb *si-* 'do, become':

(3.18) (a) arəm ududuk təsya
arəm ududuk t-Ø-si-a
water hot PFV-3SG-become-PFV
'the water has gotten hot'

(b) awr arəm ududuk tumbutusya
    awr arəm ududuk t-mbu-t-si-a
    fire water hot PFV-3.ERG-CAUS-become-PFV
    'the fire heated the water'

In an equivalent construction with a true verb, the adjective must be interpreted as a manner adverbial, not a resulting state:

(3.19) nor yo aratək təprena
       nor yo aratək t-Ø-pre-n-a
       man DEF good PFV-3SG-die-LK-PFV
       'the man died well'

Finally, adjectives and adjectival verbs can form manner adverbials with the suffix -ndi; this is not possible with true verbs:

(3.20) (a) sapikəndi tenandukuko
           sapik-ndi t-e-n-and-kuko
           correct-ADV PFV-come-LK-PFV-2DL
           'you (DL) came correctly'

       (b) nambrin urukoranakoya nana
           nambrin u-rukor-ana-k-oya nana
           eye 3.SG-go.ashore-3SG.DAT-FR.PAST-3SG mama
           makəndi sik
           mak-ndi si-k
           bad-ADV happen-FR.PAST
           'Her eyes went ashore (scanned the shore line): mama turned out badly'

       (c) kapundi mbuturukorək
           kapu-ndi mbu-t-rukor-k
           big-ADV 3.ERG-CAUS-go.ashore-FR.PAST
           'they brought them ashore with a lot of effort'

       (d) tareŋgon yowa was nuŋgo makəndi ende
           tareŋgon yowa was nuŋgo mak-ndi ende
           eagle DIST breath very bad-ADV like this
           mbukaratarək
           mbu-karat-ar-k
           3.ERG-breathe-PROG-FR.PAST
           'that eagle was breathing like this, very badly'

3.4 Quantifiers

The numeral system of Kopar is fundamentally constructed upon the anatomy of a human being. It is built on three bases, base five, base ten and base twenty. Numbers one to five in both the Kopar and Wongan dialects are as follows:

	Kopar	Wongan
'one'	ombe/mbiona/mbatep	
'two'	kombar(i)	kombri
'three'	kereməŋ	kereməŋgo
''four'	sanandək	sanandəko
''five'	tambək	

'One' and 'five' are identical in the two dialects. Among these, all but 'four' are cognate with their equivalents in Yimas, but unlike there, where 'one' through 'four' inflect for noun class and only the base 'five' is invariable, all are invariable in Kopar. The numeral 'two' can be pronounced with or without the final /i/ in the Kopar dialect. It is interesting to note that both 'four' and 'five' in the Kopar dialect have the form of adjectival verbs, with a final suffix -k. The Yimas cognate *tam* 'five' attests to the ancestral form before this accretion.

For 'six' through 'nine', 'five' is the base:

	Kopar	Wongan
'six'	tambək mbatepanda	tambək mbatepand
'seven'	tambək koɲjiranda	tambək koɲjirand
'eight'	tambək kereməŋganda	tambək keremagad
'nine'	tambək sanandəkanda	tambək sanandəkand

The final suffix apparent on these forms is the comitative postposition Kopar *nda*, Wongan *nd* 'together with', so each means 'five together with one, two, etc'. All are transparent except for 'seven' in which the component for 'two' *koɲjir* is suppletive from the basic numeral for 'two' *kompar(i)*. Note that denasalization applies in the Wongan dialect form for 'eight', but not in the Kopar dialect.

The word for 'ten' is *aytaporək* in the Kopar dialect and *aytapor* without the *-k* suffix in the Wongan dialect. This is the base for all numerals from 'eleven' through 'nineteen', simply joining 'one' through 'nine' to 'ten' as the base with the comitative postposition *nda*; I will just give Kopar dialect forms from here:

'eleven'	*aytaporək mbatepanda*
'twelve'	*aytaporək koɲjiranda*
'fifteen'	*aytaporək tambəkanda*
'eighteen'	*aytaporək tambəkanda keremənganda*
'nineteen'	*aytaporək tambəkanda sanandəkanda*

Of course, no one used these complex numerals any longer at the time of fieldwork given the moribund state of the language, if they ever really did. Tok Pisin numerals were invariably employed. These forms were elicited, but speakers were able to supply them without too much difficulty.

The numeral 'twenty' is also invariable and the base for still higher numerals. It is *pwoyn*, a word which means 'male, man' and is cognate with Murik *pwin*, Yimas *pan-mal* and Kanda *pon-do*, all of which mean 'male, man'. Hence, 'twenty' is a man, a whole man of ten fingers and ten toes which equal twenty. 'Twenty' can be simply *pwoyn*, but more commonly it is expressed as *pwoyn mbatep/mbiona* 'one man'. From 'twenty-one' to 'thirty-nine', the numerals are *pwoyn* plus 'one, two, three, up to nineteen':

'twenty-one'	*pwoyn*	*mbatepanda*
'thirty-nine'	*pwoyn*	*aytaporəkanda tambəkanda sanandəkanda*

'Forty' is *pwoyn kompar(i)* 'two men', and 'forty-one' to 'fifty-nine' would again be 'one' to 'nineteen' coordinated to that by comitative *-nda*, while 'sixty' would be *pwoyn keremən* 'three men'. Of course, no one now ever uses such higher numerals, and it is very unlikely anyone ever did, though the generative nature of the system makes them possible. In precontact times, without money and consumer goods, no one had the need to count to these levels; 'one' to 'ten' was probably more than sufficient.

There are three other quantifiers in Kopar: *waɲja* 'some, few', *awtok* 'many' and *bisi ~ bisi bisi* 'all'; *awtok* 'many' again has the *-k* suffix of adjectival verbs. In addition, one of the variant forms for 'one' *ombe* can function very much like an indefinite article such as English *a*; indefinite articles are rare in Papuan languages, but one is also found in the neighboring Watam language. Here are some examples of these quantifiers and the indefinite article:

(3.21) (a) nor waɲja təŋgiona
nor waɲja t-ŋgi-o-n-a
man few PFV-3PC-go-LK-PFV
'a few men have left'

(b) nəmbren   awtok   mbunambratundukoya
    nəmbren   awtok   mbu-nambrat-ndək-oya
    pig   many   3.ERG-spear-NR.PAST-3SG
    'he speared many pigs'

(c) mbuna   nəmandəpak   bisibisi
    mbu-na   nəmandəpak   bisibisi
    3PL-POSS   woman.PL   all
    mbuturutukududu
    mbu-t-ru-t-k-undundu
    3.ERG-CAUS-shoot-APPL-FR.PAST-3PC
    'they (PC) killed all of their (PL) wives'

(d) ndeme   nana   inda   ombe   tumbusya
    nde-ome   nana   inda   ombe   t-mbu-si-a
    how-wh   mama   house   INDEF   PFV-3.ERG-make-PFV
    'how did mama build a house?'

(e) nor   ombe   [punuarək   nəmandək   nda]
    nor   ombe   [punu-ar-k   nəmandək   nda]
    man   INDEF   work.sago-PROG-FR.PAST woman   COM
    'a man who was working sago with his wife'

(f) marikerəka   nəmandək   ombe   nasamaytaŋgoya
    ma-riker-ka   nəmandək   ombe   na-samayt-aŋg-oya
    ITR.IMP-get.up-2SG woman   INDEF   1SG.ERG-see-PRES-1SG
    'get up! I see a woman'

Quantification by *bisi* ~ *bisi bisi* 'all' need not expressed as a direct modifier within a NP. It commonly appears as an adverbial directly before the verb:

(3.22) mbukanamatukondu   bisi
        mbu-kanama-t-k-ondu   bisi
        3.ERG-chew.betelnut-APPL-FR.PAST-3PL   all
        ŋgikandəksek
        ŋgi-kandək-se-k
        3PL-sleep-NIGHT-FR.PAST
        'they (PL) chewed betelnut and they all slept that night'

And *ombe* has a reduplicated form *obeb* (note denasalization) meaning 'only' and a derived form *ambe* 'another', which in turn has specialized usage of its reduplicated form *abeb* (again denasalization), meaning 'one and (an)other':

(3.23) ambisen aymbor kwarik abeb   nana aymbor kwarik
       ambisen aymbor kwarik ambemb nana aymbor kwarik
       daughter hearth side one   mother hearth side
       abeb
       ambemb
       other
       'the daughter's hearth on one side and the mother's hearth on the other side'

3.5 Pronouns

Kopar has a very rich system of personal pronouns, most of which, in keeping with its typological profile as a head marking language, are bound pronominals on the verb for the indication of core grammatical relations and will therefore be treated in Chapter 5 on verbal morphology. Across all pronominals, Kopar distinguishes three persons and four numbers, singular, dual, paucal (a few from three to about seven) and plural (many, more than seven). The independent personal pronouns of the Kopar dialect are listed in Table 1:

Table 1: Kopar Independent Pronouns.

	SG	DL	PC	PL
1	ma	ke	paŋgə	e
2	mi	ko	ŋgu	o
3	mu	mbə	məŋgə	mbu

The independent pronouns in the Wongan dialect are the same, except that the third singular is *mə*. There are some obvious morphological formatives in these pronouns, for example, a third person stem in *m-*. This goes back to an old near distal deictic stem in other languages of the Lower Sepik-Ramu family, but has been re-analyzed as a pronoun base in Kopar. In addition, there is a paucal formative *ŋg* that goes back to Proto-Lower Sepik (Foley 2022) and, for non-third persons, a dual formative *k-*. Finally, as seen in the dual and plural, there is an ablaut alternation, *e* versus *o* for first and second person respectively. An association of a front vowel with first person non-singular and a back vowel with second person non-singular is widespread in Lower Sepik-Ramu languages and indeed even in other unrelated languages nearby, such as the Torricelli language, Mambuwan. (Foley 2017a). An interesting point to note is that *e* 'we (PL)' and *o* 'you (PL)' are homophonous with the verb roots *e-* 'come' and *o-* 'go' respectively'; surely, there is some deictic semantics of direction toward us and direction toward you behind this homophony.

The interrogative pronouns or *wh*-words collected in the corpus are:

'who'   SG: *menome*, DL: *mbame(na)*, PC: *meŋgome ~ meŋge*
'what'   *ara (moran)-(ome(na))*
         what (thing)-(wh)
'why'   *ara ŋga*
'where' *ndok-ome(na)* or *ndak-amena*
'how'   *nde-(o)me(na)*

All of these forms save 'why' can contain a formative *-ome(na)* (the initial /o/ is commonly lost following a vowel, and the final /a/ is often dropped), which is clearly the information question component parallel to English *wh-*, though it is facultative with 'what' and 'how'. 'Who' is made up of this and most likely a variant of the third person singular pronoun *ma*, so means 'he/she-which' (this is more transparent in the Murik cognate *mən-amena* 'who', where *mən-* is clearly the third person singular pronoun). The plural form for 'who' was not collected, but it is probably similarly formed on the respective third person pronoun base for number plus *ome*, as in the cognate Murik forms: SG: *mən-amena*, DL: *məndəb-amena*, PC: *məŋgəŋ-amena*, PL: *mənduŋ-amena* (note the form *-amena* is preserved in Kopar in one of the alternative words for 'where', though the possibility that this is a Murik loan cannot be excluded). Interestingly, in the corpus there is also a local person inflected form for 'who', the second person singular as in *mi meme* 2SG who.2SG < *mi* 2SG + *ome* 'wh' 'who are you?'; it is very likely there are other local person inflected forms in the language, but these were not collected. 'What' is *ara*, cognate with the Yimas word for 'what' *wara*; this often occurs in combination with the optional *wh*-formative *–ome(na)* and the word for 'thing', so 'what-thing-wh'. 'What' *ara (moran)*, as in many languages, also functions as an indefinite pronoun 'something' in Kopar. 'Why' is literally 'for what', a common crosslinguistic equivalent, witnessed for example by French *pourquoi* and Tok Pisin *bilong wanem*. 'Where' is again *-ome* plus a component *ndok-*, whose origins are unclear, though it too has a cognate in Yimas *ntuk-nti* 'how much, many' or an alternative form where /a/ replaces /o/, though this could be a Murik loan. There is yet another root for 'where' *ndata* that can function independently or as a verb root and is inflected as such:

(3.24) (a) *ndataromena*
ndata-ar-ome-naya
where-PROG-*wh*-2SG
'where are you (SG) going?' (literally 'you (SG) are where-ing?')

(b) *ndata   makandəksemekə*
ndata   ma-kandək-se-ome-okə
where   ITR.IMP-sleep-NIGHT-*wh*-1PC
'where shall we (PC) sleep?'

Finally, 'how' is composed of *nd-* plus a bound version of the postposition *e* 'like' plus the optional *wh*-formative *–ome(na)*.

3.6 Deictics

The deictic system remains one of the less well understood aspects of Kopar grammar. While they are well represented in the narrative texts, their functions and distribution are not always clear. There is a binary contrast between a proximal deictic 'this' and a distal one 'that', but they do not have the same morphological behavior, unlike their English translations. Like equivalents in Murik and Yimas, the proximal deictic inflects for number and for all four number contrasts as in the personal pronouns:

SG: *ena ~ enana*
DL: *embaya ~ enambe*
PC: *eŋgaya ~ enaŋge*
PL: *embwaya ~ enambwe*

The final *-ya* in the non-singular forms is sometimes dropped (compare 3.25c with 3.25f). What meaning difference there may be between the variants for each number is unknown. Some examples of the use of the proximal deictic:

(3.25) (a) *nor   kombari   embaya*
nor   kombari   embaya
man   two   PROX.DL
'these (DL) two men'

(b) *enambwe   ŋga   numotənda   naŋgrin*
enambwe   ŋga   numotənda   naŋgrin
PROX.PL   DAT   man.PL   spear

ŋgerakotək
ŋgi-e-ra-kot-k
3PL-come-stay-carry-FR.PAST
'the men brought the spears for these (PL)'

(c) emba mbutəmek
emba mbu-təme-k
PROX.DL 3.ERG-tell-FR.PAST
'he told these (DL)'

(d) muna enamb naok muna
mu-na ena-mb na-o-k mu-na
3SG-POSS PROX.SG-OBL 3SG-go-FR.PAST 3SG-POSS
nuno enamb
nuno ena-mb
thought PROX.SG-OBL
'he went because of this (SG) (concern) of his, his thought about this (SG)'

(e) nda yan enana mbukamendakəto
nda yan enana mbu-kame-anda-k-to
and papa PROX.SG 3.ERG-call.out-3DL.DAT-FR.PAST-DEP
'and this (SG) father called out to them (DL)'

(f) embaya inda tiŋgi
embaya inda tiŋgi
PROX.DL house behind
'these (DL) are behind the house'

(g) nor ombe enana umbiyara
nor ombe enana u-mbi-e-ar-aŋg
man one PROX.SG 3SG-AGAIN-come-PROG-PRES
'this (SG) one guy is coming again'

(h) mina ŋga enamb mbusaranaŋənaya
mi-na ŋga ena-mb mbu-sara-n-aŋg-naya
2SG-POSS DAT PROX.SG-OBL 3.ERG=1-report-LK-PRES-2SG
'I tell you about this (SG)'

The distal deictic contrasts with the proximal in that it seems to be invariable, i.e. lacks inflection for number, as no number inflected forms were found in the corpus, even if modifying non-singular head nouns (note the overt numeral 'two' in the NP of (3.26c); compare particularly (3.25a) with (3.26c). This could simply be lacunae in the data, as it is possible even to use the singular form of the proximal deictic with nouns with non-singular number, as most nouns are not overtly inflected for number, their number being determined by a pronominal verbal agreement affix, if present, otherwise indeteminate. However, both the distal and the proximal deictics inflect for the full four numbers in the most closely related language Murik (Schmidt 1953), although only the proximal deictic base is clearly cognate. The form of the distal deictic is *yowa*, and here are some examples of its use:

(3.26) (a) *indan mbunak yowa ende*
indan mbu-na-k yowa ende
house 3PL-POSS-NE DIST like this
ŋgirarorok
ŋgi-ra-ar-oro-k
3PL-stay-PROG-EXT-FR.PAST
'they stay here in that house of theirs (PL)'

(b) *sen yowa ŋgara kusambonaka*
sen yowa ŋgara ku-sambo-ana-ka
son DIST first TR.IMP-put-3SG.DAT-2SG
'put that son of hers first'

(c) *rikam kombari yowa kukatirambiko*
rikam kombari yowa ku-katira-mbi-oko
bamboo two DIST TR.IMP-cut.off-IM.FUT-2DL
'you (DL) should cut off those two bamboo (shoots)'

(d) *Wak yowa mbukinimbukududu*
Wak yowa mbu-kinimb-k-undundu
PN DIST 3.ERG-fasten-FR.PAST-3PC
'they (PC) tied that Wak up'

The distal deictic *yowa* also appears in a contracted form *yo*, which functions very much like the definite determiner in English and is very common in the narrative texts; I will gloss it as DEF. Determiners are relatively rare in Papuan langauges, though they are an areal feature of the Madang area, and Kopar has both an indefinite (examples 3.21d-f) and a definite determiner, like its distantly related neighbor Watam. This determiner use of contracted *yowa* to *yo* has clearly led to the determiner being a part of speech in its own right, as it can co-occur with an overt deictic (3.27a) Here are examples of its use (note invariable *yo* with an overtly plural head noun in (3.27d):

(3.27) (a) tareŋgo yo enana [ŋgaməmanaŋgəmbaya]
tareŋgo yo enana [ŋga-mə-ma-n-aŋg-mbaya]
eagle DEF PROX.SG INV-eat-DUR-LK-PRES-DL
'this eagle that keeps on eating us (DL)'

(b) nəmbren yo petəndəkoya
nəmbren yo petəndək-oya
pig DEF black-3SG
'the pig is black'

(c) nanan yo irormandək nanan yo mora kanda
nanan yo iror-mandək nanan yo mora kanda
mama DEF tree-female mama DEF thing sick
sisirik
si-siri-k
feel-MORNING-FR.PAST
'the mama, the tree spirit mama, felt sick in the morning'

(d) mayndəpak yo kiŋgep
mayndəpak yo kiŋgep
male.PL DEF ladder
mbuturaporaposarorokondu
mbu-t-rapo+rapo-sa-ar-oro-k-ondu
3.ERG-COM-run.RED-IN-PROG-EXT-FR.PAST-3PL
'the men kept running in with a ladder'

(e) Wak ŋga tra-ndək yo pək niŋgom
Wak ŋga tra-ndək yo pək ni-ŋga-o-am
PN DAT dance-PURP DEF bed put.inside-go.inside-go-DETR
ndək yo moran ŋgitadabarorok
ndək yo moran ŋgi-tandamba-ar-oro-k
PURP DEF thing 3PL-prepare-PROG-EXT-FR.PAST
'they (PL) kept on preparing food for Wak, for a dance, for a bed to go inside'

moran 'thing' is commonly understood as a contracted stand in for *mu-moran* eat-thing 'food'.

(f) naŋgun munak yo
naŋgun mu-na-k yo
skin 3SG-POSS-NE DEF

        ukusaməmanakoya
        u-kusa-am-ma-ana-k-oya
        3SG-come.out-DETR-DUR-3SG.DAT-FR.PAST-3SG
        'the skin of his was appearing'

    (g) nor   yo   ndesa   ma   bagariondukoya
        nor   yo   ndesa   ma   mbaŋgari-onduk-oya
        man   DEF   today   1SG   kill-FUT-1SG
        'I will kill the guy today'

The distal deictic *yowa* seems generally to be used attributively; I have few examples of its heading a noun phrase. One is in a fixed expression, *yowa namo* 'that's all' and the other in (3.28), but note the emphatic particle *ya* in this usage which may correlate with its presence in variants of the proximal deictics:

(3.28) ma-na-k   yowa   ya   dabagarike
        ma-na-k   yowa   ya   nda-mbaŋgari-oke
        1SG-POSS-NE   DIST   EMPH   NOW-kill-1DL
        'now let us (DL) kill that (one) of mine'

Otherwise, when heading an NP, distal deixis is expressed by the third person singular pronoun *mu* (Wongan *mə*):

(3.29) (a) mu   nor   kuŋgoparikoya
            mu   nor   kuŋgoparik-oya
            he/that   man   long-3SG
            'he/that is a tall man'

       (b) ambisen   mu   mbu-tu-mur-ana-k
            ambisen   mu   mbu-t-mur-ana-k
            daughter   3SG   3.ERG-COM-fear-3SG.DAT-FR.PAST
            'the daughter took fright of that'

There are two place deictics indentified in the corpus, *ende ~ aynde* 'here' and *munde* 'there' (the latter tentatively) and one manner deictic, *ende ~ endeŋ* 'so, like this, in such a manner':

(3.30) (a) məŋgə   aynde   ŋgiraraŋgiya
            məŋgə   aynde   ŋgi-ra-ar-aŋg-iya
            3PC   here   3PC-stay-PROG-PRES-PC
            'they (PC) are (staying) here'

(b) *munde   tran   yo   kwarik   ena*
    munde   tra-n   yo   kwarik   ena
    there   dance-NMLZ   DEF   side   PROX.SG
    'the dance there (on) this (SG) side'

(c) *ende   ŋgisiarorokəto*
    ende   ŋgi-si-ar-oro-k-to
    like this   3PL-do-PROG-EXT-FR.PAST-DEP
    'they (PL) kept doing it like this and'

3.7 Postpositions

Kopar like many Papuan languages has verb final basic constituent order and a right-headed phrasal typology. In alignment with this typology, Kopar has postpositions instead of prepositions. The postpositions of Kopar fall into two types: first, a smallish set that are phrasal enclitics and even occasionally have become suffixed to their complements and express more abstract grammatical case-like notions, like dative 'to' or 'for', comitative 'along with', and second, a somewhat larger set that denotes more concrete locational notions like 'inside', 'behind', 'on top of', etc.

The genitive suffix *-na* POSS attaches to both personal pronouns and nouns to indicate possessors:

(3.31) (a) *karəkarək   mana   nuŋgo   aratəkoya*
           karəkarək   ma-na   nuŋgo   aratək-oya
           pillow   1SG-POSS   very   good-3SG
           'my pillow is very good'

       (b) *məndən   kumuka   miɲjir*
           məndən   ku-mə-ka   miɲjir
           feces   TR.IMP-eat-2SG urine
           kumuka   mina   ŋaɲjenak
           ku-mə-ka   mi-na   ŋaɲje-na-k
           TR.IMP-eat-2SG 2SG-POSS   child-POSS-NE
           'eat the feces and drink the urine of your child!'

       (c) *Matawrna   nəmbən   budukundudu*
           Matawr-na   nəmbən   mbu-ndə-k-undundu
           PN of clan-POSS   garamut.drum   3.ERG-hear-FR.PAST-3PC
           'they (PC) heard Matawr's garamut'

(d) *muna   sesena   moran*
    mu-na   sese-na   moran
    3SG-POSS   grandfather-POSS   thing
    'something of his grandfather's'

(e) *sesena   inin   ŋga   utisikoya*
    sese-na   inin   ŋga   u-ti-si-k-oya
    grandfather-POSS   stone   DAT   3SG-REFL-do-FR.PAST-3SG
    urak
    u-ra-k
    3SG-stay-FR.PAST
    'he thought that grandfather's stone was still there'

The dative postposition has multiple functions, and one difference across these is whether it occurs with the genitive suffix. When it and comitative *nda* have a pronominal complement, they always require the genitive suffix, but not elsewhere:

(3.32) (a) *muna   ŋga   ruaŋ   nasaytəndukoya*
           mu-na   ŋga   ruaŋ   na-sayt-ndək-oya
           3SG-POSS   DAT   coconut   1SG.ERG-pick-NR.PAST-1SG
           'I picked a coconut for her'

   (b) *nor   puruŋ   nəmandək   ŋga   mbutukamaŋgoya*
       nor   puruŋ   nəmandək   ŋga   mbu-t-kam-aŋg-oya
       man   betelnut   woman   DAT   3.ERG-CAUS-arrive-PRES-3SG
       'the man gives betelnut to the woman'

The possessive suffix is never found in other uses of the dative postposition, such as to indicate infinitives:

(3.33) (a) *pipikambu   ŋga   o   ndana*
           pipikambu   ŋga   o   nda-na
           fish.with.hook   DAT   go   PROHIB-2SG
           'don't go fishing'

   (b) *pipikambu   ŋga   riya*
       pipikambu   ŋga   ri-oya
       fish.with.hook   DAT   DES-1SG
       'I want to fish'

Here are examples of other case-like postpositions like comitative *nda* 'with' (homophonous with 'and') and purposive *ndək* 'for':

(3.34) (a) *pwap yowa arəm nda akak*
pwap yowa arəm nda aka-k
trunk DIST water COM float-FR.PAST
'the trunk (of the tree) floated with the water'

(b) *ma kona nda pipikambu ŋga*
ma ko-na nda pipikambu ŋga
1SG 2DL-POSS COM fish.with.hook DAT
maondəkənaya
ma-o-ndək-naya
1SG-go-NR.PAST-1SG
'I went to fish with you (DL)'

(c) *naŋgep indadək nakotaraŋgoya*
naŋgep inda-ndək na-kot-ar-aŋg-oya
post house-PURP 1SG.ERG-carry-PROG-PRES-1SG
'I'm carrying a post for a house'

The other postpositions express being at, in, on, around, etc a location. These never occur with the genitive suffix. Some of these include:

su(r)	'in, on, into'	*payndəp*	'beside'
tiŋgi	'behind'	*ŋgara*	'in front of'
kaŋgarap	'above'	*tawmb*	'below'

Examples of their usage in a sentence:

(3.35) (a) *tamənd nani sur nakariaŋgoya*
tamənd nani sur na-kari-aŋg-oya
fish pot inside 1SG.ERG-put-PRES-1SG
'I put the fish into the pot'

(b) *nor iror kaŋgarap təndasana*
nor iror kaŋgarap t-Ø-ndasa-n-a
man tree above PFV-3SG-sit-LK-PFV
'the man has sat on top of the tree/wood'

(c) aymbor   payndəp   kandəksirikoya
    aymbor   payndəp   kandək-siri-k-oya
    hearth   beside   sleep-MORNING-FR.PAST-3SG
    'in the morning she slept by the side of the hearth'

There are also two words, *asawari* 'close' and *patukuri* 'far', which are not postpositions, but rather exhibit morphosyntactic properties of nouns by occurring with typical nominal markers:

(3.36) (a) *mbiraposa   asawari   ndək*
           mbi-rapo-sa asawari   ndək
           3DL-run-IN   close   PURP
           'they (DL) ran into (a place) close by'

    (b) *patukurimba   moran   mbutundatək*
        patukuri-mba   moran   mbu-t-ndat-k
        far-OBL   thing   3.ERG-CAUS-know-FR.PAST
        'she understood the issue from far away'

3.8 Temporals

Kopar has two different types of words expressing temporal notions. The first of these are day counters. Kopar, like Yimas and many other Papuan languages, has the interesting way of reckoning time from the point 'today' Janus-like in both directions, so that one day in the NR.PAST ('yesterday') and one day in the future ('tomorrow') is denoted by the same term *rari* '1.day.removed'. The actual day referred by *rari* is determined by the choice of tense on the verb, near past versus future. The day counters collected are:

ndesa   'today'
rari   '1.day.removed'
(rari) uta   '2.days.removed'

It is possible and quite common, though, to specify 'yesterday' by adding the near past tense suffix to *rari* '1.day.removed' or *uta* '2.days.removed', yielding, for *rarindək* 'yesterday' or *utandək* 'day before yesterday'; a parallel usage with the future tense meaning 'tomorrow' or 'day after tomorrow' is not found.

Another set of temporal words expresses the parts of the day:

tumbunan 'morning' (5am-9am)
sinan 'middle of the day' (9am-4pm)
wakənan 'late afternoon' (4pm-7pm)
rakamdan 'night' (7pm-5am)

Ending in an optional /n/ these have the formal structure of nouns (see section 4.1, though the Wongan dialect equivalents lack the final /n/), but any of these can also be suffixed with the near past tense suffix *-ndək* (minus the final /n/), a feature of verbs, to express a time of the day yesterday: *tumbunandək* 'yesterday morning'. Furthermore, Kopar verbs can have incorporated temporal suffixes corresponding to each of these temporal words:

(3.37) (a) *mu upipikambusirindukoya*
mu u-pipikambu-siri-ndək-oya
3SG 3SG-fish.with.hook-MORNING-NR.PAST-3SG
'he fished in the morning'

(b) *nor kapu yo mu ok*
nor kapu yo mu o-k
man big DEF 3G go-FR.PAST
ukirisek
u-kiri-se-k
3SG-copulate-NIGHT-FR.PAST
'the mature man, he went and had sex during the night'

(c) *kakandək ŋgari sora kamək*
kakandək ŋgari sora kam-k
older.sister DAT shell.type give-FR.PAST
nda ŋgiramak
nda ŋgi-ra-ma-k
and 3PL-stay-AFTERNOON-FR.PAST
ŋgipunumak
ŋgi-punu-ma-k
3PL-work.sago-AFTERNOON-FR.PAST
'he gave the older sister a shell and they stayed and worked sago in the afternoon'

A few other words denoting temporal notions are:

patemba	'a long time ago'
sinan	'day'
karep	'month, moon'
mberi	'year'

The last three of these are clearly syntactically nouns; unlike the noun-like temporal words which express parts of the day that can take a near past tense suffix, a diagnostic of verbs, these cannot (*sinan* can do so only in the sense of 'middle, hottest, part of the day', never in the sense of a 'day', i.e. a period of twenty-four hours). The word *patemba* is derived from a root *pate* meaning 'before' and a locative/temporal suffix -*mba*.

3.9 Conjunctions

Kopar has only one native conjunction, *nda* 'and', transparently linked to and homophonous with the comitative postposition *nda* 'with', that is used to conjoin both NPs and clauses. Examples of its use follow:

(3.38) (a) *mbə nana nda ambise nda kayn mbatep*
mbə nana nda ambise nda kayn mbatep
3DL mother and daughter and canoe one
mbitəkandaŋgakəndi
mbi-t-kanda-ŋga-k-ndi
3DL-COM-jump-go.into-FR.PAST-ADV
mbiosirik
mbi-o-siri-k
3DL-go-go.downriver-FR.PAST
'while they (DL), a mother and daughter, jumped together into one canoe, they (DL) went downriver'

(b) *nana e ŋga usianarəkoya*
nana e ŋga u-si-ana-ar-k-oya
mama come DAT 3SG-do-3SG.DAT-PROG-FR.PAST-3SG
nda akənjim o mamrayn nda pamba
nda akən-jim o mamrayn nda pamba
and sun-cloud thunder and just

            was   nda   pamba   nda
            was   nda   pamba   nda
            wind   and   just   and
            '(when) mama wanted to come, (there will be) rain clouds and
            thunder and wind and'

    (c)   nana   nda   yaya   mbirikerəkəmbaya
            nana   nda   yaya   mbi-riker-k-mbaya
            mama   and   papa   3DL-get.up-FR.PAST-3DL
            'mama and papa got up'

    (d)   moran   mbupukorak   nda   nəmandək   raway
            moran   mbu-pukora-k   nda   nəmandək   raway
            thing   3.ERG-give.food-FR.PAST and   woman   self
            ambise   sik   muna   ŋga
            ambise   si-k   mu-na   ŋga
            daughter become-FR.PAST 3SG-POSS DAT
            'she gave her food and the woman herself
            became a daughter to her'

    (e)   nda   mbu   o   nda   e   nda
            nda   mbu   o   nda   e   nda
            and   3PL   go   and   come   and
            ŋgisikəŋgaya
            ŋgi-si-k-ŋgaya
            3PL-do-FR.PAST-3PL
            'and they (PL) kept going and coming'

Otherwise, Kopar, like many Papuan languages, makes heavy use of simple juxtaposition or suffixes on dependent verbs in clause chaining constructions to link clauses together, and this will be treated in Chapter 7. However, the Tok Pisin conjunctions *na* 'and' especially and *orait* 'and then' are also now common in Kopar discourse. Here are examples of these from narrative texts:

(3.39) (a) na   mayn   ukamenakoya
            na   mayn   u-kame-ana-k-oya
            and   husband   3SG-call.out-3.SG.DAT-FR.PAST-3SG
            na   mu   esek
            na   mu   ese-k
            and   3SG   bring-FR.PAST
            'and (her) husband called out to her and she brought (it)'

(b) orait   na   sokayn   mbukwaymbukondu
    orait   na   sokayn   mbu-kwaymb-k-ondu
    and.then and   tobacco 3.ERG-divide-FR.PAST-3PL
    'and then they divided up the tobacco'

3.10 Interjections

Given the limited amount of data, only a few interjections were recorded. They are:

awo	'yes'
kaya	'no'
o	'oh'
e	'eh'
eya	'hey'

Chapter 4
Nouns and noun phrases

4.1 Nouns

Compared to its sister languages like Yimas and Kanda, nouns are morphologically simple in Kopar. With the exception of nouns with human referents, there are no obligatory inflections for most nouns, not even number, in spite of the rich number distinctions in pronouns. One distinguishing feature of a subset of nouns, those ending in /n/, is that they often lose this final consonant in combination with a postposition or even when functioning alone as the head of a NP. Consider these examples with *indan* 'house' and *numbon* 'sago soup':

(4.1) (a) *tərəma   inda   sur*
          t-Ø-rəmə-a   inda   sur
          PFV-3SG-stand-PFV house inside
          'he stood up inside the house'

      (b) *naŋgep   mbutendukoya   inda   undi   ŋga*
          naŋgep   mbu-t-e-ndək-oya   inda   undi   ŋga
          post   3.ERG-CAUS-come-NR.PAST-3SG house build DAT
          'he brought a post to build a house'

      (c) *nəmandək   numbo   təma*
          nəmandək   numbo   t-Ø-m-a
          woman   sago.jelly PFV-3SG-eat-PFV
          'the woman has eaten sago jelly'

This final /n/ in such words goes back to a fossilized singular marker still apparent on cognates in Murik and other Lower Sepik languages. Interestingly, some nouns ending in /n/ when elicited in the Kopar dialect are always given with the final /n/, but not so in the Wongan dialect:

	Kopar	Wongan
'pig'	nəmbren	nəmbre
'house'	indan	inda
'canoe'	kayn	kay
'path'	porakayn	porakay

Seemingly related to this are nouns which have variable singular forms like:

'man, male, husband'	*mayn*	*mayndək*
'woman, wife'	*nəman*	*nəmandək*
'older.same.sex.sibling'	*kakan*	*kakandək*
'younger.sister ♀'	*itəman*	*itəmandək*

Here the final /n/ is replaced by a suffix *–ndək,* clearly homophonous with the near past tense suffix *-ndək,* but, as these are simple nouns, it obviously is not functioning like that here. This spread of *-ndək* seems quite promiscuous in Kopar, sometimes for no clear function, as with certain adjectival verbs:

petəndək 'dark, black'   *patendək* 'old'
patəndək 'heavy'

The word for 'black, dark' also has a shortened form without the suffix, *pet,* and the word for 'heavy' arguably is derived from *pat* 'stone', an Austronesian loan (note the use of 'stone' to denote a unit of weight even in English). The root *pate* in 'old' is also found in the temporal word *pate-mba* 'a long time ago', where *-mba* is a locative/temporal suffix to a root *pate* 'before'.

There is a likely explanation for these forms. There is an adjectival and attributivizing suffix in closely related Murik *–(a)rogo,* which is clearly cognate with Kopar *-ndək;* the sound changes are regular and the Murik vowel is accounted for by vowel harmony to the final appended singular marker *-o* (the plural form of the Murik suffix is actually *–(a)rək,* even more transparently cognate to Kopar *-ndək).* This Murik suffix is ubiquitous on adjectives derived from nouns and is also optionally added to nouns with human referents, just as in Kopar: *nəma(rogo)* 'woman, female', *awr* 'fire', *awr-arogo* 'warm'. The ultimate origin of this *-ndək,* *-(a)rogo* is almost certainly the purposive postposition *ndək,* but it has become bound and clearly semantically bleached in its usage here.

The final /n/, however, does appear to function productively as a nominalizing suffix which can be added to verbs to derive nouns. Some common examples:

Verb		Noun	
kata-	'to speak'	*katan*	'speech'
sara-	'to report, narrate'	*saran*	'a story'
tay-	'to try, attempt'	*tayn*	'ability'
tra-	'to dance'	*tran*	'a dance'

si-	'to do, become'	sin	'an action, event'
kame-	'to call out'	kamen	'a call, shout, cry'

The only inflectional grammatical category that nouns can take is number: nouns can be marked for plural. For nouns with human referents, plural marking is close to obligatory except in combination with quantifiers. Plural marking is typically by a suffix *–pak ~ -rəmbak*, though there can be some irregularities and some nouns have alternate plurals:

	SG	PL
'person, man'	nor	normot/norəmbak/numotanda
'husband, man, male, spouse'	mayn/mayndək	maynem/mayndəpak/mayndəpari
'woman, wife, spouse"	nəman(d)/nəmandək	nəmandəpak
'older.same.sex.sibling'	kakan/kakandək	kakarəmbak
'younger.brother ♂'	ram(ay)	ramarəmbak/ramtaŋgar
'younger.sister. ♀'	itəman/itəmandək	itəmarəmbak/itəmataŋgar
'cross.sex.sibling'	maman	mamarəmbak/mamataŋgar

In older forms of the language, as witnessed by its sister Murik (Foley 2022), nouns were probably inflected for four numbers, singular, dual, paucal and plural, just like pronouns. The plurals in *-ŋgar* actually go back to older paucal forms, and those with *–rəmbak* most probably to older duals. The suffix *-mot* in *nor-mot* 'people, men' is the old plural for this root, cognate with its Murik equivalent. The other, suppletive, plural for *nor* 'person, man' actually is composed of *numot* 'village' plus the comitative postposition *nda,* so 'with the village', i.e, co-villagers'.

Count nouns with non-human referents can optionally occur with a pluralizing morpheme *okumbi*; this strictly means plural, i.e. 'more than one' and is quite distinct from any numeral or the quantifier *awtok* 'many':

	SG	PL	
'dog'	uren	uren	okumbi
'tooth'	asirap	asirap	okumbi
'hand'	napar	napar	okumbi
'star'	kinaŋ	kinaŋ	okumbi

okumbi is juxtaposed to NPs and follows deictics and the determiner *yo: karan yo okumbi* head DEF PL 'the heads'. While pluralization of such nouns with *okumbi* is acceptable; in reality it is not commonly found. In ongoing Kopar discourse such nouns are normally simply left unmarked for number.

4.2 Noun compounds

Compounds consisting of two nouns are attested in Kopar and are formed by simply juxtaposing the noun stems. A few examples are:

> akən-jim 'rain cloud' < akən 'sun' + jim 'cloud'
> nəmandək-ŋaɲjen 'girl' < nəmandək 'woman, wife' + ŋaɲjen 'child'
> uren-mayn 'dog-man' < uren 'dog' + mayn 'man'
> nimbre-kame-n 'pig grunt' < nimbre(n) 'pig' + kame-n call-NMLZ 'a call'
> kunden-kayn 'canoe made of black palm bark' < kunden 'black palm bark'
>   + kayn 'canoe'
> uren-ŋaɲjen 'puppy' < uren 'dog' + ŋaɲjen 'child'
> nəma-kusep 'female crab' < nəma 'woman' + kusep 'crab'

Verb plus noun compounds are also possible as in:

> kandək-nambrin 'sleepiness' < kandək- 'to sleep' + nambrin 'eye'
> mu-moran 'food' < mə- 'to eat (with vowel harmony) + moran 'thing'

4.3 Noun phrases

Noun phrases tend not to be greatly elaborated in Kopar; NPs with more than four constituents were never encountered, even under elicitation. With the exception of possessive pronouns modifiers always follow their head nouns.

4.3.1 Possession in noun phrases

Here we will only be concerned with possessive constructions within NPs; for clause level predication of possession, like 'I have a house' see section 6.2.4. Within NPs, possessors are always marked by the genitive suffix -na, which attaches itself to either pronominal or nominal heads. Possessive pronouns are simply the independent pronouns to which -na is suffixed, as listed in Table 2:

Table 2: Kopar Possessive Pronouns.

	SG	DL	PC	PL
1	ma-na	ke-na	paŋgə-na	e-na
2	mi-na	ko-na	ŋgu-na	o-na
3	mu-na	mbə-na	məŋgə-na	mbu-na

These possessive pronouns can either precede or follow their head nouns, and like many Papuan languages, there is no distinction between alienable and inalienable possession:

(4.2) (a) ma-na   uren   (b) karan ma-na
          1SG-POSS dog   head   1SG-POSS
          'my dog'   'my head'

When postnominal modifiers occur in a noun phrase, prenominal possessive pronouns are highly favored, postnominal position is awkward, if not outright ungrammatical:

(4.3) (a) ma-na   uren kapu   (b) ??uren kapu ma-na
          1SG-POSS dog   big   dog   big   1SG-POSS
          'my big dog'   'my big dog'

Postnominal, but not prenominal, possessors commonly occur in a second form ending in *-k*. This suffix seems to function much like *-ne* in English possessive paradigms like *my/mine, thy/thine*; hence I gloss it as NE. So next to (4.3a), we find (4.4b):

(4.4) (a) uren ma-na   (b) uren ma-na-k
          dog   1SG-POSS   dog   1SG-POSS-NE
          'my dog'   'my dog'

Literally (4.3b) means 'dog mine' or perhaps better colloquial English 'dog of mine', and while it may look like an equational sentence in Kopar with a zero copula, i.e. 'the dog is mine' (Kopar lacks a copula), that analysis is not plausible because possessors with *-k* NE occur with extraposed possessors in non-equational sentences:

(4.5) məndən kumuka   miɲjir
       məndən ku-mə-ka   miɲjir
       feces   TR.IMP-eat-2SG urine

kumuka	*mina*	*ŋaɲjenak*
ku-mə-ka	mi-na	ŋaɲje-na-k
TR.IMP-eat-2SG	2SG-POSS	child-POSS-NE

'eat the feces and drink the urine of your child!'

In this example the possessor of *məndən* 'feces' and *miɲjir* 'urine' has been extraposed and hence is self-standing and requires the suffix *-k* NE. Note as well that it is impossible for *mi-na* 2SG-POSS, the possessor of *ŋaɲje(n)* 'child', to follow it when suffixed with *-k*: *ŋaɲje-na-k mi-na. Also postposed possessors with *-k* NE occur before the determiner *yo* DEF, indicating that they can also be constituents within a noun phrase:

(4.6) *mayn*   *mbə-na-k*   *yo*
       husband 3DL-POSS-NE DEF
       'the husband of theirs (DL)'

4.3.2 Postnominal modification in noun phrases

A fully expanded NP would have the following structure:

(4.7) POSS + N + Adjective + Numeral + Adjectival Verb + Deictic/Determiner

But as mentioned above, noun phrases composed of more than three constituents are not attested in Kopar; the above formula is a composite derived from attested combinations. Fully expanded noun phrases were never encountered even in elicitation and would certainly seem very awkward to Kopar speakers. Even those with three were mainly elicited, rarely found in the narrated texts, though easily provided. Examples of expanded NPs include:

(4.8) (a) *nor* *tambək*
          man   five
          'five men'

    (b) *indan* *kombar* *yowa*
          house   two   DIST
          'those two houses'

    (c) *maɲjikap* *keremən* *nuŋgo* *patəndək*
          netbag   three   very   heavy
          'three very heavy netbags'

(d) indan   kapu kombari
    house big   two
    'two big houses'

(e) ma-na   ŋaɲjen keremən
    1SG-POSS child   three
    'my three children'

As (4.8a) demonstrates, nouns with human referents are not pluralized when enumerated or quantified. In comparing (4.8c) with (4.8d) note that the adjective *kapu* 'big' and the adjectival verb *patəndək* 'heavy' have different placements with respect to the numeral, before and after respectively. It is not known if alternative orders are possible, but these examples were what was obtained on elicitation, so would most likely be the unmarked orders. These data again are in line with the findings of section 3.3 concerning the formal distinction between the adjective and adjectival verbs.

Postposed possessors with *-k* rarely occur if adjectives, adjectival verbs or numerals are present and are most common with the deictics or the determiner *yo* (4.9a, b), but a construction with an adjectival verb is attested in one example (4.9c) with an unexpected word order:

(4.9) (a) naŋgun   mu-na-k   yo
         skin   3SG-POSS-NE DEF
         'his skin'

     (b) indan   mbunak   yowa ende ŋgirarorok
         indan   mbu-na-k   yowa ende ŋgi-ra-ar-oro-k
         house 3PL-POSS-NE DIST here 3PL-stay-PROG-EXT-FR.PAST
         'they (PL) stayed here (in) that house of theirs (PL)'

     (c) karəkarək   manak   nuŋgo aratək
         karəkarək   ma-na-k   nuŋgo aratək
         pillow   1SG-POSS-NE   very   good
         'a very good pillow of mine'

4.4 Relative clauses

No relative clauses were obtained in elicitation, but they do occur sporadically in the narrative texts. Because all four narrative texts are legends, they were narrated in the far past tense, so the examples of relative clauses are mostly limited to that tense. It is not known therefore if there are any structural differences for relative clauses according to tense, but there is one example of a relative clause in present tense (4.12a) and it shows no morphosyntactic differences. Kopar appears to allow both externally headed and internally headed relative clauses, and those with external heads need not be embedded, but can be adoined (Hale 1976). (4.10a) is an example of an adjoined relative clause with an external head, while (4.10b) is one that is internally headed:

(4.10) (a) kar   kwarakotəmbiko
           kar   ku-wa-ra-kot-mbi-ko
           feast   TR.IMP-go-stay-carry-IM.FUT-2DL
           [mbu   mbusambokondu]
           [mbu   mbu-sambo-k-ondu]
           3PL   3.ERG-leave-FR.PAST-3PL
           'you (DL) take the feast that they (PL) left'

      (b) mbu-na   numbon
          mbu-na   numbon
          3PL-POSS   sago.jelly
          [nəmandəpak   mumoran   sik]   yo
          [nəmandəpak   mu-moran   si-k]   yo
          woman.PL   eat-thing make-FR.PAST   DEF
          'their sago jelly, the food the women cooked' ('cook' here is literally 'make thing to eat')

In (4.10a) the external head is *kar* 'feast', but the relative clause modifying it does not form a constituent with it, rather it is postposed after the main verb; this is an adjoined relative clause (Hale 1976). In (4.10b) the internal head of the relative clause is *mu-moran* eat-thing 'thing to eat' or 'food', but the whole relative clause modifies *numbon* 'sago jelly', which is the specific food stuff which has been cooked by the women. Note the presence of the definite determiner *yo* closing off the noun phrase containing the relative clause in (4.10b). This usage is very common, but is not obligatory; compare (4.10a).

The relativized noun can be in any function in Kopar relative clauses, although relativization of subjects is the most frequent in the corpus. Examples of relativization on subjects are:

(4.11) (a) nəmandəpak [numbon tesek]   yo
             nəmandəpak [numbon t-e-se-k]   yo
             woman.PL   sago.jelly CAUS-come-NIGHT-FR.PAST DEF
             'the women who brought the sago jelly at night'

     (b) nəmandəpak [ŋgisararək]
         nəmandəpak [ŋgi-sara-ar-k]
         woman.PL   3PL-report-PROG-FR.PAST
         'women who were narrating'

     (c) numotanda [ek]   yo
         numotanda [e-k]   yo
         person.PL come-FR.PAST DEF
         'the people who came'

     (d) nor ombe [punuarək   nəmandək nda]
         nor ombe [punu-ar-k   nəmandək nda]
         man INDEF work.sago-PROG-FR.PAST woman   COM
         'a man who was working sago with (his) wife'

     (e) nor ombe [uyarək]   mbusamaytəkondu
         nor ombe [u-e-ar-k]   mbu-samayt-k-ondu
         man INDEF 3SG-come-PROG-FR.PAST 3.ERG-see-FR.PAST-3PL
         'they (PL) saw a man coming'

     (f) Wak yowa sesena   iniŋ   ŋga
         Wak yowa sese-na   iniŋ   ŋga
         PN   DIST grandfather-POSS stone DAT
         utisikoya   [urak]
         u-ti-si-k-oya   [u-ra-k]
         3SG-REFL-do-FR.PAST-3SG   3SG-stay-FR.PAST
         'that Wak thought about grandfather's stone which remained'

Note the indefinite determiner *ombe* precedes the relative clause in (4.11d, e), while the definite determiner *yo* must follow it. This indicates that the indefinite determiner *ombe* has not yet fully shed its categoriality as a numeral, as numerals occur postnominally but closer to the head than deictics, the source of the definite determiner.

Examples of objects as relativized nouns are less common and examples available involve adjoined relative clauses:

(4.12) (a) *moran   mbutundatək*
moran   mbu-t-ndat-k
thing   3.ERG-CAUS-know-FR.PAST
[*ende   ŋgi-si-ar-aŋ-əgaya*]
[ende   ŋgi-si-ar-aŋ-ŋaya]
here   3PL-do-PROG-PRES-3PL
'they (PL) understood what they (PL) were doing here'

(b) *kar   kwarakotəmbiko*
kar   ku-wa-ra-kot-mbi-ko
feast TR.IMP-go-stay-carry-IM.FUT-2DL
[*mbu   mbusambokondu*]
[mbu   mbu-sambo-k-ondu]
3PL   3.ERG-leave-FR.PAST-3PL
'you (DL) take the feast that they (PL) left'

(c) *mbə   mumoran   yo   mbuturikerəkondə*
mbə   mə-moran   yo   mbu-t-riker-k-ondə
3DL   eat-thing DEF   3.ERG-CAUS-get.up-FR.PAST-3DL
mbutupapaŋgitətak
mbu-t-papaŋgit-ta-k
3.ERG-CAUS-throw.out.(*paŋgit*- RED)-OUT-FR.PAST
'they (DL) threw out the food they (DL) had prepared'

It is also possible to have oblique constituents inside postpositional phrases as relativized head nouns; again these are adjoined:

(4.13) (a) *bagabaga   sur   mbindasak*
bagabaga   sur   mbi-ndasa-k
platform   on   3DL-sit-FR.PAST
[*moran   mbimarək*]
[moran   mbi-mə-ar-k]
thing   3DL-eat-PROG-FR.PAST
'they (DL) sat on the platform where they (DL) were eating the food'

(b) *ŋgiokəto   [arəm kiarək]   inimor   sur*
ŋgi-o-k-to   [arəm ki-ar-k]   inimor   sur
3PC-go-FR.PAST-DEP water bathe-PROG-FR.PAST hole   in
'they (PC) went to the (water)hole where they were bathing'

Note the relative clause precedes the head noun in (4.13b).

Finally, it is even possible to relativize the possessor, although here a pronominal trace in the relative clause seems required:

(4.14) nəmandək   [ambisen   munak
        nəmandək   [ambisen   mu-na-k
        woman   daughter   3SG-POSS-NE
        upararisenak]   yo
        u-pa-rari-se-ana-k]   yo
        3SG-STILL-cry-NIGHT-3SG.DAT-FR.PAST   DEF
        'the woman whose daughter was still crying for her during the night'

There is one probable example of a headless relative clause in the corpus as well:

(4.15) [prek]   yo   mbukinimbukududu
        [pre-k]   yo   mbu-kinimb-k-undundu
        die-FR.PAST DEF   3.ERG-fasten-FR.PAST-3PC
        'they wrapped up the dead'

Chapter 5
Verbal morphology

This chapter represents the core of Kopar grammar, for, as befitting its polysynthetic status, it is in the complexities of verbal inflection and derivation that the real genius of the language reveals itself. Kopar verbal morphology is very rich, expressing tense, aspect, mood, time of day, person and number of the subject or object and recipient, causative, reflexive, accompaniment, manner, direction, etc. It is also diverse, sometimes exhibiting an ergative alignment for person-number marking and sometimes an accusative alignment, depending on the tense-aspect-mood marking of the verb. In those tense-aspect categories that show ergative alignment, there is a secondary split, person-number marking now being determined by a direct-inverse system sensitive to the relative ranking of participants along an Animacy Hierarchy (Dixon 1979; Silverstein 1976). The complexity in the expression of core arguments according to their person and number makes it very difficult to provide a coherent gloss in term of grammatical function for them. Consider the third person singular suffix -*oya*, indicating the person-number of an intransitive subject in (5.1a) but a transitive subject in (5.1b) or the first person singular suffix -*naya* indicating an intransitive subject in (5.1c), but a transitive object in (5.1d):

(5.1) (a) nor   umundukoya
          nor   u-mə-ndək-oya
          man 3SG-eat-NR.PAST-3SG
          'the man ate'

     (b) nor   nəmandəpak   mbusamaytundukoya
          nor   nəmandəpak   mbu-samayt-ndək-oya
          man woman.PL   3.ERG-see-NR.PAST-3SG
          'the man saw the women'

     (c) mamaraŋənaya
          ma-mə-ar-aŋ-naya
          1SG-eat-PROG-PRES-1SG
          'I am eating'

     (d) mbu   ŋasamaytaŋənaya
          mbu   ŋa-samayt-aŋ-naya
          3PL   INV-see-PRES-1SG
          'they (PL) see me'

Because the matching of bound pronominals to the functions of core arguments is so varied, I will only provide a gloss for their grammatical function in those cases where it is consistent, such as for the prefixes for first singular and third person transitive subjects of realized events, which I will gloss as ergative (ERG); the rest will simply be specified for person and number.

5.1 Transitivity

As befits a partially ergative language, the notion of transitivity and a contrast between transitive and intransitive verbs is crucial to understanding the inflectional patterns of Kopar. Ditransitive verbs have variable inflection, sometimes like transitive, verbs, so that the recipient argument of ditransitive verbs is sometimes just marked by the dative postposition *ŋga* and not crossreferenced by a bound pronominal, and other times indicated by a bound dative pronominal. In the tense-aspect-mood paradigms that work on an accusative alignment, intransitive and transitive verbs inflect alike, for instance in the immediate future:

(5.2) (a) *mi   karəmbina*
 mi   kar-mbi-ona
 2SG walk-IM.FUT-2SG
 'you (SG) should walk'

 (b) *mi   kayn   pakambina*
 mi   kayn   paka-mbi-ona
 2SG canoe carve-IM.FUT-2SG
 'you (SG) will carve a canoe'

Example (5.2a) contains an intransitive verb and (5.2b) a transitive one, but in each case the person-number inflection is the same: a second person singular subject is indicated by the suffix *-ona*. Now consider the corresponding forms in the near past tense:

(5.3) (a) *mi   ikarəndəkənaya*
 mi   i-kar-ndək-naya
 2SG 2-walk-NR.PAST-2SG
 'you (SG) walked'

(b) mi   kayn   ipakandukona
    mi   kayn   i-paka-ndək-ona
    2SG   canoe 2-carve-NR.PAST-2SG
    'you (SG) carved a canoe'

In (5.3) the markers for second person singular are distinct: a suffix *-naya* for intransitive verbs and *-ona* for transitive verbs. This is an ergative alignment system: *-ona* is the second singular ergative suffix for transitive subjects, while *-naya* is the nominative one for intransitive subjects and transitive objects. Whether a transitive verb inflects on an accusative or ergative basis is determined by the tense-aspect-mood inflection of the verb. To generalize here, tense-aspect-mood categories that denote unrealized events inflect on an accusative basis, while those which express realized ones do so on an ergative basis. This is a not unexpected split along tense-aspect-mood lines from what has been documented in other languages (Dixon 1994).

Verb roots are divided into intransitive and transitive classes; even verbs of similar meaning can be categorized differently:

(5.4) (a) mu   uri   ŋga   urwaŋgoya
          mu   uri   ŋga   u-ru-aŋg-oya
          3SG   crocodile   DAT   3SG-shoot-PRES-3SG
          'he shoots at the crocodile'

      (b) mu   uri   mbunambrataŋgoya
          mu   uri   mbu-nambrat-aŋg-oya
          3SG   crocodile   3.ERG-spear-PRES-3SG
          'he spears the crocodile'

      (c) saran   mbikatandəkəmbaya
          sara-n   mbi-kata-ndək-mbaya
          report-NMLZ   3DL-speak-NR.PAST-3DL
          'they (DL) reported something'

This is a cognate object verb and hence inflected intransitively; see section 6.1.2.2.

      (d) nor   mbutəmendukondə
          nor   mbu-təme-ndək-ondə
          man   3.ERG-tell-NR.PAST-3DL
          'they (DL) told the man'

There are some verb roots identified that are both transitive and intransitive; one is *mə-* 'eat, drink, smoke':

(5.5) (a) ma (mora) mamənsdəkənaya
ma (mora) ma-mə-ndək-naya
1SG thing 1SG-eat-NR.PAST-1SG
'I ate'

(b) ma tamənd namundukoya
ma tamənd na-mə-ndək-oya
1SG fish 1SG.ERG-eat-NR.PAST-1SG
'I ate fish'

In the intransitive (5.5a), the dummy cognate object *mora(n)* 'thing' may or may not appear, but it does not affect the transitivity of the verb, which remains intransitive (verbs with cognate objects are generally intransitive, see section 6.1.2.2) and inflects as such in the near past tense with a prefix *ma-* and a suffix *-naya* both indicating first person singular. (An alert reader will have noticed the homophonous suffix *-naya* for second singular in (5.3a). What distinguishes first from second person here are the different prefixes *ma-* 1SG versus *i-* 2. As we shall see, overall this is unusual in Kopar, as person-number distinctions are normally indicated by suffixes.) The transitive verb in (5.5b) has an overt specific object, and the clause is formally transitive, so the verb is inflected transitively with a first person singular ergative prefix *na-* and suffix *-oya* for the transitive subject.

5.2 The pronominal affix agreement systems for core arguments

5.2.1 The accusatively aligned system

As mentioned above, the bound pronominals of some tense and mood categories inflect according to an accusative alignment, that is, the bound pronominals denote transitive and intransitive subjects. The tenses and moods that have this accusative alignment of bound pronominals denote unrealized events and include the immediate future, the future, negative, the desiderative ('want') and its negation ('don't want'), the prohibitive, i.e the negative imperative, and the imperative. The inflectional pattern of the imperative is slightly different from the others, but it still essentially patterns accusatively. In these tenses and moods, all bound pronominals are suffixal; the following Table 3 presents the underlying forms for these subject marking pronominal suffixes:

5.2 The pronominal affix agreement systems for core arguments — 69

Table 3: Bound Subject Pronominals in Unrealized Tense-Moods.

	SG	DL	PC	PL
1	-oya	-oke	-okə	-onde
2	-ona	-oko	-onduku	-ondo
3	-onda	-ondə	-undundu	-ondu

The initial /o/ ~ /u/ of these suffixes always elides after a suffix ending in a vowel or semivowel and sometimes elsewhere (5.6e). Here is the inflectional paradigm for *kar-* 'walk' in the immediate future, marked by the suffix *-mbi* (the epenthetic /ə/ between the verb root and the immediate tense suffix is omitted here for transparency of the paradigm):

	SG	DL	PC	PL
1	kar-mbi-ya	kar-mbi-ke	kar-mbi-kə	kar-bi-de
2	kar-mbi-na	kar-mbi-ko	kar-bi-duku	kar-bi-do
3	kar-bi-da	kar-bi-də	kar-bi-dudu	kar-bi-du

The same set of suffixes would mark the subjects of a transitive verb like *nambrat-* 'spear', as this paradigm is accusatively aligned. Direct object marking with bound pronominals never occurs with these subject markers. Note that the suffixes with prenasalized stops following the immediate future tense suffix *-mbi* trigger mutual denasalization by rule (2.12), causing both to denasalize and resulting in a sequence of syllables with plain voiced stops; in the case of the third paucal suffix, both of its underlying prenasalized voiced stops are denasalized. The final /ə/ of *-okə* 1PC is also commonly omitted.

Here are a few examples of the use of the suffixes of Table 3 with other unrealized moods:

(5.6) (a) ma arəm kiri ŋga riya
        ma arəm ki-ri ŋga ri-oya
        1SG water bathe-DOWN DAT DES-1SG
        'I want to bathe'

    (b) ma ndaya ndə ŋga
        ma nda-oya ndə ŋga
        1SG don't.want-1SG hear DAT
        'I don't want to listen'

(c) mu   rari   ndonduk   kaynda
    mu   rari   ndə-onduk   kay-onda
    3SG   1.day.removed   hear-FUT   NEG-3SG
    'he won't listen tomorrow'

(d) mi   arəm   kiri   ndana
    mi   arəm   ki-ri   nda-ona
    2SG   water   bathe-DOWN   PROHIB-2SG
    'don't bathe!'

(e) ruaŋ   saytondukonda
    ruaŋ   sayt-onduk-onda
    coconut   pick-FUT-3SG
    'he will pick a coconut'

(f) ma   rari   samaytondukoya
    ma   rari   samayt-onduk-oya
    1SG   1.day.removed   see-FUT-1SG
    'I will see him tomorrow'

(g) awr   rari   siondukona
    awr   rari   si-onduk-ona
    fire   1.day.removed   make-FUT-2SG
    'you (SG) will light a fire tomorrow'

(h) ma   nəmand   ŋgatondukoya   sen   nda
    ma   nəmand   ŋga-t-o-onduk-oya   sen   nda
    1SG   woman   FIRST-CAUS-go-FUT-1SG   son   COM
    'I will take the woman with her son first'

Imperatives are slightly more complex but pattern accusatively also (see 5.3.4.2). In addition to a suffix indicating the number of the second person addressee subject, they take a prefix indicting whether the verb is transitive or intransitive:

(5.7) (a) *ma-rəmə-ka*
          ma-rəmə-ka
          ITR.IMP-stand-2SG
          'you (SG) stand up!'

      (b) *tamənd   kumuka*
          tamənd   ku-mə-ka
          fish   TR.IMP-eat-2SG
          'you (SG) eat the fish'

5.2.2 The ergatively aligned system

The ergatively aligned system of bound pronominals is much more complex than the accusatively aligned system, and this is essentially for two reasons. Firstly, while there are two sets of bound pronominals, an ergative set and a nominative one for transitive verbs, the bound pronominals of both sets consist of prefixal and suffixal components, and the subject and object compete for the same prefixal and suffixal slots. Both a direct-inverse system and person neutralization operate in transitive verbs, and these determine which grammatical function, subject or object, gets realized in these slots. The tenses that inflect according to an ergative alignment are those for realized events, the far past, the near past and the present. The perfective also has bound pronominals according to an ergative alignment, but its inflectional patterns are slightly different and will be treated in section 5.3.1.4.

In presenting this complex system of ergatively aligned pronominal marking for core arguments, subject and object, it is easier to start with the simplest pattern, that of intransitive verbs. The following Table 4 presents the underlying forms for the bound affixes for all person-number combinations for intransitive verbs:

Table 4: Bound Subject (Nominative) Pronominals for Realized Tenses in Intransitive Verbs.

	SG	DL	PC	PL
1	ma-....-naya	i-....-mbake	i-....-iya	i-....-mbwade
2	i-....-naya	i-....-mbako	i-....-iya	i-....-mbwado
3	u-....-oya	mbi-....-mbaya	ŋgi-....-iya	ŋgi-....-ŋgaya

There are a number of recurring features to take note of in these affixes. First, there is a clear division between the local persons, first and second, and the non-local third person. Only the non-local third persons indicate their number by bound pronominal prefixes. For the first and second local persons, all contrasts collapse to *ma-* [-addressee] versus *i-* [+addressee]. The language, like all Lower Sepik languages, lacks an inclusive-exclusive opposition, so only the first person singular is [-addressee]; all non-singular first person forms can include the addressee and so occur with *i-* [+addressee]. The prefixal system can be summarized as Table 5:

Table 5: Analysis of Kopar Pronominal Prefixes for Intransitive Verbs in Realized Tenses.

	[-addressee]		ma-	
[+local]				
	[+addressee]		i-	
[-local]		u-	mbi-	ŋgi-
		SG	DL	PL

There is also an alternative prefix form *na-* for third singular, most commonly used with the motion verbs *e-* 'come' and *o-* 'go'. Compensating for the denuded system of these bound pronominal prefixes for local persons, there is a full set of bound pronominal suffixes, as seen in Table 4. The contrast between first and second person in non-singular number except for paucal is marked by the vowel: mid front vowel /e/ for first person and mid back vowel /o/ for second person. In addition, the dual and plural suffixes for first and second seem to be built on the corresponding third person independent pronoun plus a linking vowel /a/ plus the pronominal suffixes from Table 3 for the paradigm for the unrealized tenses and moods: *-mbako* 2DL < *mbə* 3DL + *a* + *-oko* 2DL, *-mbwade* < *mbu* 3PL + *a* + *-onde* 1PL (with loss of /o/ and denasalization of /nd / in *-nde,* but not mutual denasalization of /mb/ of *-mbu,* most likely due to paradigm pressure from the dual forms that have /mb/ initials). Also note that the paucal suffix never distinguishes person, being invariably *–iya* PC, so the first and second person paucal forms are homophonous. The first and second person singular suffixes are also homophonous, but the inflected verbs are always distinguishable by the prefixes *ma-* [-addressee] versus *i-* [+addressee]. The initial /n/ of these first and second person singular suffixes is commonly elided following the present tense indicated by a suffix *-aŋg,* as is the prenasalization of initial prenasalized voiced stops in the dual and plural suffixes (surprisingly mutual denasalization does not apply here, although it does occur optionally in the case of third plural suffix). Finally, the suffixes of the non-local third persons except paucal seem largely to be derived from the initial segment of the corresponding prefix to which is appended a sequence *–(a)ya*. This formative *–(a)ya* is also found in all the paucal suffixes and in the first and second person singular suffixes. It seems to be omissible in many contexts for all of these suffixes except for third person singular, in which it is always present. This *–(a)ya* also sometimes replaces the local person formatives in the dual and plural, so that *-mbaya* can mark any person in the dual number, and, for example, the first plural is sometimes realized as *-mbwaya* instead of the usual *-mbwade.* Also, it should be mentioned that in ongoing Kopar discourse like narrative texts, verbs often do not occur with both the prefixes and the suffixes: the third singular often loses the prefixal component *u-,* while non-singular third person subjects are commonly marked just with the prefixal component, and verbs suffixed with the dependent verb suffixes always lose their pronominal suffixes (see section 7.2.3). The [-addressee] prefix *i-* also is sometimes omitted.

I present here paradigms of the intransitive verb *kar-* 'walk' in both present and near past tenses (again epenthetic /ə/ is omitted for the transparency of the paradigm):

(5.8) (a) present tense: *-aŋg*

	SG	DL	PC	PL
1	*ma-kar-aŋg-aya*	*i-kar-aŋg-bake*	*i-kar-aŋg-iya*	*i-kar-aŋg-bwade*
2	*i-kar-aŋg-aya*	*i-kar-aŋg-bako*	*i-kar-aŋg-iya*	*i-kar-aŋg-bwado*
3	*u-kar-aŋg-oya*	*mbi-kar-aŋg-baya*	*ŋgi-kar-aŋg-iya*	*ŋgi-kar-aŋg-gaya*
				or *ŋgi-kar-ag-gaya*

(b) near past tense: *-ndək*

	SG	DL
1	*ma-kar-ndək-naya*	*i-kar-ndək-mbake*
2	*i-kar-ndək-naya*	*i-kar-ndək-mbako*
3	*u-kar-nduk-oya*	*mbi-kar-ndək-mbaya*

	PC	PL
1	*i-kar-ndik-iya*	*i-kar-nduk-mbwade*
2	*i-kar-ndik-iya*	*i-kar-nduk-mbwado*
3	*ŋgi-kar-ndik-iya*	*ŋgi-kar-ndək-ŋgaya*

These forms are from the Kopar dialect. The Wongan dialect diverges in reinforcing the distinction in person between first and second, but neutralizing that in number between dual and plural. Here is the paradigm of *kar-* 'walk' in near past tense with durative aspect marker *-mana* 'was walking':

(5.9)

	SG	DL
1	*ma-kar-mana-ndək-naya*	*i-kar-mana-ndək-mbaya*
2	*n-kar-mana-ndək-naya*	*n-kar-mana-ndək-mbaya*
3	*kar-mana-nduk-oya*	*mbə-kar-mana-ndək-mbaya*

	PC	PL
1	*i-kar-mana-ndik-iya*	*i-kar-mana-ndək-mbaya*
2	*n-kar-mana-ndik-iya*	*n-kar-mana-ndək-mbaya*
3	*i-kar-mana-ndik-iya*	*ŋgi-kar-mana-ndək-ŋgaya*
		or *i-kar-mana-ndək-ŋgaya*

There are a number of differences between the two dialects. The Wongan dialect has a prefix *n-* that is used with all second person subjects, so that in the non-singular first and second person are distinguished by *i-* versus *n-* respectively. Hence while first and second paucal are homophonous in the Kopar dialect, they are distinct in the Wongan dialect. But the prefix *i-* extends into third person paucal and plural (optionally replacing *ŋgi-* in the plural), so that the contrast in those numbers is *n-* second person versus *i-* first and third person, meaning that first and third paucal are now homophonous. Also, except for third person plural, the suffixes in dual and plural number have become invariable like that of paucal number, with *-mbaya* 3DL, generalizing across all persons; this generalization of *-mbaya* in the dual forms, as mentioned earlier, is also a possible option in the Kopar dialect. Finally, the prefix *u-* 3SG is lost.

Transitive verbs are much more complicated because there are now two core arguments, subject and object, competing for the bound pronominal slots. Transitive verbs introduce a second set of bound pronominal affixes, also a mix of prefixes and suffixes; the ergative bound pronominal series in the Kopar dialect are presented in Table 6:

Table 6: Bound Subject (Ergative) Pronominals for Realized Tenses in Transitive Verbs.

	SG	DL	PC	PL
1	na-....-oya	i-....-oke	i-... -ok(ə)	i-....-onde
2	i-....-ona	i-....-oko	i-....-onduku	i-....-ondo
3	mbu-....-oya	mbu-....-ondə	mbu-....-undundu	mbu-....-ondu

The only difference between these and those of the Wongan dialect is the form for third paucal, *i-. . . .-oya*, resulting from a loss of person distinctions in prefixes for paucal number and the replacement of the distinct third paucal suffix *-undundu* by the corresponding singular suffix *-oya*. Note the identity of the suffixal components of these forms in Table 6 to the suffixes of Table 3. The suffixes containing prenasalized voiced stops again lose their prenasalization following the present tense suffix *-aŋ*. Commonly, this will only affect the pronominal suffix, leaving the nasal in the present tense suffix intact, though sometimes this will trigger instead mutual denasalization by rule (2.12): *mbu-nambrat-ar-ag-odə* 3.ERG-spear-PROG-PRES-3DL 'they are spearing it'. The /u/ of the third person ergative prefix *mbu-* is deleted before a verb root beginning in /u/ or /o/ as well as /w/ such as *wa-* 'go'. Finally, the vowels of the second and third paucal suffixes are sometimes realized as /ə/ and the final vowel sometimes elided, perhaps on analogy with the first person form.

A quick comparison of the bound pronominal forms of Tables 4 and 6 reveals an obvious problem. There is only one prefixal position and only one suffixal position for bound pronominals available in Kopar verbs, and both sets contain prefixes and suffixes marking a certain person-number combination for a particular grammatical function like transitive subjects for the ergative series. For transitive verbs, those with two core arguments, subject and objects, both compete for these prefixal and suffixal positions, and what determines the winner is essentially the Animacy Hierarchy. The Animacy Hierarchy in Kopar ranks local persons, first and second, over non-local third persons, and within the local persons, it ranks first person over second. Kopar operates a direct-inverse system for transitive verbs which is overlaid on the contrasting sets of pronominals for grammatical functions of Tables 4 and 6. Whenever a higher ranked transitive subject acts on a lower ranked transitive object, we find the direct inflectional pattern, while when a lower ranked transitive subject acts on a higher ranked object, we get the inverse inflectional pattern. When both participants are of equal rank in Kopar, that is non-local third persons, we get neither, a neutral pattern.

5.2.2.1 Neutral inflectional pattern: Non-local person acts on non-local person

This is the simplest inflectional pattern. When both participants are of equal rank, that is non-local third person, generally the subject is marked by a bound pronominal affix, in this case obviously from the ergative series for transitive subjects:

(5.10) (a) mbu   indan   mbundimanagodu
           mbu   indan   mbu-undi-ma-n-aŋg-ondu
           3PL   house   3.ERG-build-DUR-LK-PRES-3PL
           'they (PL) are working at building a house'

     (b) nor   nəmbren   mbunambratundukoya
         nor   nəmbren   mbu-nambrat-ndək-oya
         man   pig   3.ERG-spear-NR.PAST-3SG
         'the man speared a pig'

     (c) nəmandəpak   uren   mbutumanəŋaŋgududu
         nəmandəpak   uren   mbu-tumanəŋ-aŋg-undundu
         woman.PL   dog   3.ERG-hit-PRES-3PC
         'the women (PC) are hitting the dog'

     (d) Matawr-na   nəmbən   budukududu
         Matawr-na   nəmbən   mbu-ndə-k-undundu
         PN.clan-POSS   garamut.drum   3.ERG-hear-FR.PAST-3PC
         'they (PC) heard the garamut drum of Matawr'

(e) mayndəpak   mbuturarorokondə
    mayndəpak   mbu-t-ra-ar-oro-k-ondə
    husband.PL   3.ERG-COM-stay-PROG-EXT-FR.PAST-3DL
    'they (DL) stayed with their husbands'

While this is the general pattern, there are other options. It is possible to indicate the person and number of a non-singular subject with a prefix drawn from the nominative series of Table 4 and employ a suffix from ergative series of Table 6 to indicate its number (5.11a) or to simply mark a transitive verb with a prefix from Table 4 and omit the suffix entirely (5.11b) or use the third singular suffix from Table 4 to indicate the object (5.11c):

(5.11) (a) mbisamaytaraŋgodə
           mbi-samayt-ar-aŋg-ondə
           3DL-see-PROG-PRES-3DL
           'they (DL) are looking at it'

       (b) moran   ŋgitadabarorok
           moran   ŋgi-tandamba-ar-oro-k
           thing   3PL-prepare-PROG-EXT-FR.PAST
           'they (PL) were preparing things'

       (c) Wak   ombe   ŋgara   ŋgitenaŋgoya
           Wak   ombe   ŋgara   ŋgi-t-e-n-aŋg-oya
           PN   one   first   3PC-CAUS-come-LK-PRES-3
           'they (PC) brought Wak alone first'

As mentioned earlier, an overt noun phrase recipient of ditransitive verbs is an oblique phrase marked with the dative postposition ŋga. However, ditransitive verbs with third person recipients have two possible inflections patterns. They can inflect like simple monotransitive verbs, with the same inflectional variants:

(5.12) (a) nəmandək   ŋga   puruŋ   mbutukamaŋgodə
           nəmandək   ŋga   puruŋ   mbu-t-kam-aŋg-ondə
           woman   DAT   betelnut 3.ERG-CAUS-arrive-PRES-3DL
           'they (DL) give betelnut to the woman'

       (b) mbu   nəmandəpak   ŋga   puruŋ   ŋgitakamək
           mbu   nəmandəpak   ŋga   puruŋ   ŋgi-t-kam-k
           3PL   woman.PL   DAT   betelnut 3PL-CAUS-arrive-FR.PAST
           'they (PL) gave betelnut to the women'

Or they can occur with one of the dative suffixes (see section 5.2.4). The verb for 'give' can occur with or without the causative prefix *t-*; for the etymology of this formation see section (6.1.2.3):

(5.13) (a) nda itəmandək   enana
           nda itəmandək   enana
           and younger.sister ♀ PROX.SG
           mbu-kam-ana-k
           mbu-kam-ana-k
           3.ERG-give-3SG.DAT-FR.PAST
           'and (her) younger sister gave this to her'

   (b) *napar mbutukaməmanak*
       napar   mbu-t-kam-ma-ana-k
       hand   3.ERG-CAUS-arrive-DUR-3SG.DAT-FR.PAST
       'he was giving (his) hand to her'

   (c) *iniŋ   ŋgikamanak   Wak*
       iniŋ   ŋgi-kam-ana-k   Wak
       stone   3PL-give-3SG.DAT-FR.PAST PN
       'they (PL) gave the stone to Wak'

5.2.2.2 Direct inflectional pattern: Local person acts on non-local person

The direct inflectional pattern largely follows the basic neutral pattern: only subjects can be indicated by bound pronominals, those of the ergative series, in both the prefixal and suffixal positions:

(5.14) (a) *e   indan   iundimanaŋgode*
           e   indan   i-undi-ma-n-aŋg-onde
           1PL house   1-build-DUR-LK-PRES-1PL
           'we (PL) are working at building a house'

   (b) *nəmbren   imaraŋgoke*
       nəmbren   i-mə-ar-aŋg-oke
       pig   1-eat-PROG-PRES-1DL
       'we (DL) are eating pork'

   (c) *kayn   ipakandukona*
       kayn   i-paka-ndək-ona
       canoe   2-carve-NR.PAST-2SG
       'you (SG) carved a canoe'

(d) *paŋgəna   saran   indaraŋoko*
paŋgə-na   sara-n   i-ndə-ar-aŋg-oko
1PC-POSS   talk-NMLZ   2-hear-PROG-PRES-2DL
'you (DL) are listening to our (PC) story'

(e) *ikakwamanaŋok*
i-kakwa-ma-n-aŋg-okə
1-peel-DUR-LK-PRES-1PC
'we (PC) keep peeling them off'

(f) *mana   nana   ibagarindukona*
ma-na   nana   i-mbaŋgari-ndək-ona
1SG-POSS   mama   2-kill-NR.PAST-2SG
'you (SG) killed my mother'

(g) *mbu-na   ŋga   ruaŋ   isaytundukona*
mbu-na   ŋga   ruaŋ   i-sayt-ndək-ona
3PL-POSS   DAT   coconut   2-pick-NR.PAST-2SG
'you (SG) picked a coconut for them (PL)'

(i) *taward   imaraŋone*
taward   i-mə-ar-aŋg-ona-e
fish   2-eat-PROG-PRES-2SG-Q
'you are eating fish?'

The behavior of ditransitive verbs is also the same as with verbs in neutral inflection. Noun phrase recipients of ditransitive verbs are marked obliquely with the dative postposition *ŋga*, and the object of such verbs is the theme or the object transferred. Hence ditransitive verbs with a local subject and a non-local third person theme object have the direct inflectional pattern, as seen in (5.14g). The corpus provides no examples of verbs in direct inflection with non-local person dative pronominal suffixes for recipients of ditransitive verbs.

First singular ergative subject differs from the other local person-number combinations in that, just as with the nominative forms for intransitive verbs in Table 4, it has a unique prefix form *na-* rather than the ubiquitous *i-*.

(5.15) (a) *ma   nor   natumanəŋaŋgoya*
ma   nor   na-tumanəŋ-aŋg-oya
1SG   man   1SG.ERG-hit-PRES-1SG
'I hit the man'

(b) məŋgə   nasamaytundukoya
　　məŋgə   na-samayt-ndək-oya
　　3PC   1SG.ERG-see-NR.PAST-1SG
　　'I saw them (PC)'

(c) urenpətak   nakatamanaŋgoya
　　uren-pətak   na-katama-n-aŋg-oya
　　dog-morsel   1SG.ERG-taste-LK-PRES-1SG
　　'I taste dog meat'

(d) nəmandək   ŋga   puruŋ   natəkamaŋgoya
　　nəmandək   ŋga   puruŋ   na-t-kam-aŋg-oya
　　woman   DAT   betelnut 1SG.ERG-CAUS-arrive-PRES-1SG
　　'I give betelnut to the woman'

5.2.2.3 Inverse inflectional pattern: Non-local person acts on local person

The inverse inflectional pattern occurs whenever a lower ranked person acts on a higher ranked one, i.e. third person acts on first or second, or second person acts on first, and is indicated by a prefix ŋga-, which usurps the prefixal component of the bound pronominal affix. In this section I will consider the case of a non-local third person subject acting on local first or second person object; in the following section I will discuss when a second person subject acts on first person object. In the inverse inflection, due to the usurpation of the prefixal slot by the inverse prefix ŋga-, the person-number of a core argument can now only be marked by a suffix. When the subject is a non-local third person participant, that suffix slot always references the object and by the suffixal component of the nominative suffixes, those used with subjects of intransitive verbs of Table 4; hence this is clearly an ergative pattern of inflection:

(5.16) (a) nor   ŋgasamaytəndəkənaya   inda   sur
　　　　　 nor   ŋga-samayt-ndək-naya   inda   sur
　　　　　 man   INV-see-NR.PAST-1SG   house   inside
　　　　　 'the man saw me inside the house'

　　　 (b) nəmbre   ŋgakaymatəndəkənaya
　　　　　 nəmbre   ŋga-kaymat-ndək-naya
　　　　　 pig   INV-gore-NR.PAST-2SG
　　　　　 'the pig gored you (SG)'

(c) nəmandəpak ŋgasamaytarundukumbwado
    nəmandəpak ŋga-samayt-ar-ndək-mbwado
    woman.PL   INV-see-PROG-NR.PAST-2PL
    'the women were watching you (PL)'

(d) Peta ŋgu ŋgatumanəŋarangiya
    Peta ŋgu ŋga-tumanəŋ-ar-aŋg-iya
    PN   2PC INV-hit-PROG-PRES-PC
    'Peter is hitting you (PC)'

(e) ŋgatərararaŋgənaya
    ŋga-t-ra-ar-aŋg-naya
    INV-COM-stay-PROG-PRES-1SG
    'she looks after me'

(f) gadanambrataŋgəna
    ŋga-nda-nambrat-aŋg-naya
    INV-NOW-spear-PRES-2SG
    'he (will) spear you now'

Note the dropping of the formative –(a)ya in (5.16f).

Ditransitive verbs (with some exceptions, see section 6.1.2.2) with a third person subject and a first or second person recipient behave just as they do in direct inflection. As the recipient is marked obliquely with the dative postposition *ŋga,* the object of the verb is the theme or the thing transferred, another third person participant, hence the verb has the neutral inflectional pattern for non-local third person acting on non-local third person (5.17a).

(5.17) məŋgə mana   ŋga uren ŋgitenagududu
       məŋgə ma-na   ŋga uren ŋgi-t-e-n-aŋg-undundu
       3PC   1SG-POSS DAT dog 3PC-CAUS-come-LK-PRES-3PC
       'they (PC) bring me a dog'

These examples demonstrate the general system of inverse inflection, available for any combination of a non-local third person acting on a local first or second person. However, there seems to be other options for local objects. Instead of invariant inverse marker *ŋga-,* special inverse ergative pronominal prefixes, as in its sister language Murik (Foley 2016), can be used with local direct object arguments: *mbi- ~ mba-.* It is also possible to use *mbu-,* normally 3.ERG, instead of *ŋga-* INV in this context of a non-local person subject acting on a local object:

(5.18) (a) *tareŋgo enana mbiməmanaŋgəmbaya*
tareŋgo enana mbi-mə-ma-n-aŋg-mbaya
eagle   PROX.SG 3.ERG/1.OBJ-eat-DUR-LK-PRES-DL
'this eagle keeps eating us (DL)'

(b) *mbə ŋgu mbatumanəŋaŋgiya*
mbə ŋgu mba-tumanəŋ-aŋg-iya
3DL 2PC 3.ERG/2.OBJ-hit-PRES-PC
'they (DL) are hitting you (PC)'

(c) *nəmandəpak mbukatatakwatarandide*
nəmandəpak mbu-kata-takwat-ar-andə-nde-e
woman.PL   3.ERG/1.OBJ-speak-lie-PROG-PFV-1PL-Q
'I think the women have been lying to us (PL)'

5.2.2.4 Inverse inflectional pattern: Local second person acts on local first person

The verbal forms for a local second person subject acting on a local first person object show more variation. It is not known whether this is due to paradigmatic breakdown and confusion resulting from the moribund state of the language or whether this variation is a stable grammatical feature of this configuration. It is a well established typological generalization that forms in which local persons act on other local persons present especial difficulties and exhibit irregularities across many languages (Heath 1997), and Kopar is in line with this observation. In the Kopar Animacy Hierarchy, first person outranks second person, so in this situation a lower ranked subject acts on a higher ranked object. This calls for the inverse inflectional paradigm: the verb takes the inverse prefix *ŋga-*, and the suffixal agreement is with the first person object, the higher ranked participant, again using a nominative suffix from Table 4, as befits this ergative inflectional pattern:

(5.19) (a) *mi ŋgasamaytəndəkənaya inda sur*
mi ŋga-samayt-ndək-naya inda sur
2SG INV-see-NR.PAST-1SG   house inside
'you (SG) saw me inside the house'

(b) *ŋgu ŋgatumanəŋarangəbake*
ŋgu ŋga-tumanəŋ-ar-aŋg-mbake
2PC INV-hit-PROG-PRES-1DL
'you (PC) are hitting us (DL)'

Note that these verb forms are ambiguous. They are identical to the inverse forms in 5.2.2.3 and without the independent pronoun (5.19a) could mean 'he/she/they are hitting me'. The independent pronouns are necessary to distinguish second person subjects from third persons subjects in these inverse forms. The person contrast between second and third person is lost in these verbs.

This ambiguity could be the source of variation in this configuration. Look back at the affixes in Table 4; note that the suffixes for first and second singular are identical: -naya. This means that the suffix on the verb in (5.19a) could be second singular, not first singular. So that in spite of the inverse prefix ŋga-, if this suffix is analyzed as second singular, the verb then agrees with the subject as in the direct forms of 5.2.2.2, rather than with the object as in the inverse forms of 5.2.2.3. This is exactly what we find in the variant forms for this configuration of second person subject acting on first person object (compare (5.20) with (5.19a):

(5.20) mi   ŋgasamaytəndəkənaya   inda   sur
       mi   ŋga-samayt-ndək-naya inda   sur
       2SG INV-see-NR.PAST-2SG   house   inside
       'you (SG) saw me/us in the house'

Note that this verb is also ambiguous, as -naya can be parsed as a first or second person singular object with a third person subject in an inverse construction 'he/they saw me/you (SG)' (see example (5.16a). Crucially, though, if -naya is taken as marking a second person subject in (5.20), the object can only be first person, as the construction is inverse, hence lower ranking subject acting on higher ranking object. But in this case, it is now the number contrast of the first person object which is lost. While the function and person of -naya in (5.19a) and (5.20) is ambiguous due to its homophony, this is not the case in other numbers of second person subjects when they appear as suffixes; they clearly mark the subject function and its person-number:

(5.21) (a) ko   ke   ŋgatumanəŋarangəbako
           ko   ke   ŋga-tumanəŋ-ar-aŋg-bako
           2DL 1DL   INV-hit-PROG-PRES-2DL
           'you (DL) are hitting us (DL)'

       (b) o   paŋgə   ŋgasamaytundukumbwado
           o   paŋgə   ŋga-samayt-ndək-mbwado
           2PL 1PC   INV-see-NR.PAST-2PL
           'you (PL) saw us (PC)'

Of course, the examples in (5.21) could also be inverse verbs with third person subjects and second person objects if the independent pronouns were omitted: 'he/they saw you (DL/PL)'. It is important to note that these second person pronominal suffixes, while indicating the transitive subject, are still formally from the nominative series of Table 4, those used for intransitive subjects and transitive objects in the inverse constructions of 5.2.2.4.

In a sense the variants of (5.20) and (5.21) give contradictory signals, as perhaps befits the problematic nature of a local person acting on another local person. The inverse prefix ŋga- indicating a lower ranking subject acting on a higher ranked object argues that for the Animacy Hierarchy second person is ranked lower than first person. On the other hand, the fact a second person subject optionally can grab the prominent suffixal position for bound pronominals in these alternative constructions suggests that second person outranks first person, like in Algonkian languages (Hockett 1966). By having a bet each way, these constructions highlight the difficulty that configurations of two local participants present. We will see this again in the following section.

There are no examples in the corpus of ditransitive verbs with second person subjects and first person recipients having inverse inflection nor can a first person recipient ever be indicated by a dative suffix of section 5.2.4. The object is a third person theme, and the inflection is direct:

(5.22) (a) *mana*   *ŋga*   *puruŋ*   *itenaŋgona*
  ma-na   ŋga   puruŋ   i-t-e-n-aŋg-ona
  1SG-POSS   DAT   betelnut   2-CAUS-come-LK-PRES-2SG
  'you (SG) bring me betelnut'

(b) *ma-na*   *ŋga*   *ruaŋ*   *iniɲjaragoduku*
  ma-na   ŋga   ruaŋ   i-niɲja-ar-aŋg-onduku
  1SG-POSS   DAT   coconut   2-send-PROG-PRES-2PC
  'you (PC) are sending me a coconut'

5.2.2.5 Impersonalization: Local first person acts on local second person

This is the second fraught configuration of local person acts on local person, and indeed it is the most troubled of all. At this point the language goes beyond inversion and opts for complete impersonalization. The first person subject is realized by the third person ergative prefix *mbu-* 3.ERG, and as such all number distinctions are neutralized. The second person object with full number specifications is indicated by the bound pronominal suffix, again drawn from the nominative series of Table 4:

(5.23) (a) ma   mbusamaytəndəkənaya
           ma   mbu-samayt-ndək-naya
           1SG 3.ERG=1-see-NR.PAST-2SG
           'I saw you (SG)'

      (b) paŋgə mbutumanəŋaraŋgubwado
          paŋgə mbu-tumanəŋ-ar-aŋg-bwado
          1PC   3.ERG=1-hit-PROG-PRES-2PL
          'we (PC) are hitting you (PL)'

      (c) ma   mi   mbe   ŋga
          ma   mi   mbe   ŋga
          1SG   2SG   give.birth DAT
          mbutenaŋgənaya
          mbu-t-e-n-aŋg-naya
          3.ERG=1-CAUS-come-LK-PRES-2SG
          'I'm bringing you to give birth'

Note that in spite of the use of *mbu-* 3.ERG to express first person subjects, these verbs unlike those in section 5.2.2.4 are not ambiguous. They cannot denote third person subjects, because that would require the inverse inflectional pattern of 5.2.2.3 and an overt inverse prefix *ŋga-*. These verbs can only be taken as having first person subjects, although all number contrasts in the first person subject is lost because of this process of impersonalization.

Ditransitive verbs with first person subjects and second person recipients usually do not exhibit impersonalization, although they can (5.24b). Again, as the object is a third person theme, the inflection is typically direct, as in (5.24a), but the corpus does provide one example of impersonalization (5.24b), in which the object is the second person recipient marked by the nominative suffix:

(5.24) (a) mina   ŋga   puruŋ   itəkamaŋgok
           mi-na   ŋga   puruŋ   i-t-kam-aŋg-okə
           2SG-POSS   DAT   betelnut 1-CAUS-arrive-PRES-1PC
           'we (PC) give you (SG) betelnut'

       (b) mina   ŋga   enamb   mbusaranaŋgənaya
           mi-na   ŋga   ena-mb   mbu-sara-n-aŋg-naya
           2SG-POSS   DAT   PROX.SG-OBL   3.ERG=1-report-LK-PRES-2SG
           'I tell you (SG) about this (SG)'

Example (5.24a) is the usual direct pattern for clauses with first person subjects and second person recipients. However, there are still other less common alternatives that were provided in elicitation sessions, though never found in the spontaneous narrative texts. Their grammatical status in unclear; they may be older patterns or may or may not be due to speaker uncertainly given the complexity of the bound pronominal system and the advanced moribund state of the language. Speakers give (5.25a, b) as equivalent:

(5.25) (a) ma   mina   ŋga   puruŋ   natəkamaŋgoya
           ma   mi-na   ŋga   puruŋ   na-t-kam-aŋg-oya
           1SG   2SG-POSS   DAT   betelnut 1SG.ERG-CAUS-arrive-PRES-1SG
           'I give you (SG) betelnut'

       (b) ma   mina   ŋga   puruŋ   mbutukamaŋgoya
           ma   mi-na   ŋga   puruŋ   mbu-t-kam-aŋg-oya
           1SG   2SG-POSS   DAT   betelnut 3.ERG=1-CAUS-arrive-PRES-?
           'I give you (SG) betelnut'

The verb of example (5.25b) in isolation without independent pronouns can also mean 'he/she gives betelnut to X'. Example (5.25a) is the expected direct inflectional pattern of a first person singular subject acting on a third person theme object, parallel to (5.24a). Example (5.25b) is problematic in that it shows impersonalization via *mbu-* 3.ERG, indicating a first person acting on a second person, in this case the recipient participant, but the suffixal agreement is indeterminate because it could either be first singular or third singular (look back at Table 6: the suffixal components of the first singular and third singular ergative are homophonous). Here is another example of the inflectional pattern of (5.25b)

(5.26) ma   mina   ŋga   ruaŋ   mbuniɲjandukoya
       ma   mi-na   ŋga   ruaŋ   mbu-niɲja-ndək-oya
       1SG   2SG-POSS   DAT   coconut   3.ERG=1-send-NR.PAST-?
       'I send you (SG) a coconut'

So, the question these examples pose is whether *-oya* is third person, so that impersonalization applies to both the prefix and suffix or whether it is the first person ergative suffix, marking the person-number of the transitive subject in spite of its impersonalization by the prefix *mbu-*. The data are inconclusive on this point, so the which of these two analyses is correct must remain indeterminate, pending any possible collection of further data.

5.2.3 Pronominal agreement in the perfective

The pronominal agreement paradigms in the perfective as a realized tense-aspect also work at least partially according to an ergative alignment, and again with a direct-inverse contrast, but the forms and distinctions involved are somewhat different from the other realized tenses, the far past, near past and present. The perfective is marked typically simultaneously by a prefix *t-* and a suffix *-andə ~ -a* (the second allomorph is found with third person and first singular in the Kopar dialect and the first elsewhere, but in the Wongan dialect *-andə* is found in all person and numbers), though it is common in the narrative texts collected to employ either only the prefix or suffix. Table 7 below provides the perfective pronominal agreement system with intransitive verbs:

Table 7: Bound Pronominal Affixes (Nominative) in the Perfective for Intransitive Verbs.

	SG	DL	PC	PL
1	ma- ~ -ma	-kike	-kəkə	-nde
2	-ana(na)	-kəko	-nduku	-ndo
3	Ø ~ na-	mbi-	ŋgi-	ŋgi-

Some of these suffixes partially overlap (minus the initial /o/) with those of Table 6. Note that prenasalization is always lost in those suffixes with prenasalized voiced stops (2PC, 1PL and 2PL); this is due to the presence of the prenasalized voiced stop /nd/ in the perfective suffix *-andə* preceding them. Here are some examples of intransitive verbs in the perfective:

(5.27) (a) *təmama*
t-ma-mə-a
PFV-1SG-eat-PFV
'I've eaten'

(b) *asawa təsiandəma*
asawa t-si-andə-ma
close PFV-do-PFV-1SG
'I'm almost done'

(c) *nor tərəma*
nor t-Ø-rəmə-a
man PFV-3SG-stand-PFV
'the man has stood up'

(d) *nəmandəpak arəm təŋgikira*
nəmandəpak arəm t-ŋgi-ki-ri-a
woman.PL water PFV-3PL-wash-DOWN-PFV
'the women have bathed'

(e) *təkandəkandəkəko*
t-kandək-andə-kəko
PFV-sleep-PFV-2DL
'you (DL) have slept'

(f) *nor ombe tena*
nor ombe t-Ø-e-n-a
man INDEF PFV-3SG-come-LK-PFV
'a man has come'

(g) *nor ombe tənaena*
nor ombe t-na-e-n-a
man INDEF PFV-3SG-come-LK-PFV
'a man has come'

(g) *təkaronandanana*
t-kar-o-n-andə-anana
PFV-walk-go-LK-PFV-2SG
'you (SG) have gone walking'

(h) *sapikindi tenandəkəko*
sapiki-ndi t-e-n-andə-kəko
good-ADV PFV-come-LK-PFV-2DL
'it's good that you (DL) have come'

The bound pronominal affixes for transitive subjects are a mixed bag. There are distinct ergative affixes only for first singular and third person, prefixes identical to the corresponding ergative prefixes of Table 6. For other person-number combinations, the same suffixes as in Table 7 are used to indicate the transitive subject, again undergoing in the relevant forms loss of prenasalization following *-andə*. Hence, the transitive perfective paradigm exhibits a split ergative-accusative alignment, seemingly according to person, ergative for first singular and third person, and accusative for all others, but actually determined formally, ergative when there is a prefix available, otherwise accusative. Table 8 presents the relevant forms:

Table 8: Bound Pronominal Affixes in the Perfective for Transitive Verbs.

	SG	DL	PC	PL
1	*na-*	*-kike*	*-kəkə*	*-nde*
2	*-ana(na)*	*-kəko*	*-nduku*	*-ndo*
3	*mbu-...-a*	*mbu-...-a* or *-ndə*	*mbu-...-a* or *-ndundu*	*mbu-...-a* or *-ndu*

Some examples to illustrate follow:

(5.28) (a) *tamənd tənama*
tamənd t-na-mə-a
fish PFV-1SG.ERG-eat-PFV
'I've eaten fish'

    (b) *tənasambota*
t-na-sambo-t-a
PFV-1SG-leave-APPL-PFV
'I've left it'

    (c) *nor təndaduduku*
nor t-ndə-andə-nduku
man PFV-hear-PFV-2PC
'you (PC) have listened to the man'

    (d) *nor nəmbren tumbusamayta*
nor nəmbren t-mbu-samayt-a
man pig PFV-3.ERG-see-PFV
'the man has seen a pig'

    (e) *indan tundiandanana*
indan t-undi-andə-anana
house PFV-build-PFV-2SG
'you (SG) have built a house'

    (f) *kayn təpakanandəkəkə*
kayn t-paka-n-andə-kəkə
canoe PFV-carve-LK-PFV-1PC
'we (PC) have carved a canoe'

5.2 The pronominal affix agreement systems for core arguments — 89

Transitive verbs in the perfective occur in the neutral (5.29a, b), direct (5,29c, d), and inverse (5.29e, f) inflectional forms:

(5.29) (a) *numotanda nəmbren tumbumadudu*
numotanda nəmbren t-mbu-mə-andə-ndu
man.PL pig PFV-3.ERG-eat-PFV-3PL
'the men have eaten pork'

(b) *nor nəmandək tumbutəmena*
nor nəmandək t-mbu-təme-n-a
man woman PFV-3.ERG-tell-PFV
'the man has told the woman'

(c) *mi awr tisiandanana*
mi awr t-si-andə-anana
2SG fire CAUS-become-PFV-2SG
'you (SG) have lit a fire'

(d) *ma nor tənasamayta*
ma nor t-na-samayt-a
1SG man PFV-1SG.ERG-see-PFV
'I have seen the man'

(e) *nor ma təŋgasamayta*
nor ma t-ŋga-samayt-a
man 1SG PFV-INV-see-PFV
'the man has seen me'

(f) *nəmandəpak mbukatatakwataradide*
nəmandəpak mbu-kata-takwat-ar-andə-nde
woman.PL 3.ERG/1.OBJ-speak-lie-PROG-PFV-1PL
'I think the women have been lying to us (PL)'

Examples (5.29e, f) are examples of inverse verb forms in the perfective in the corpus. (5.29f) is the rarer type of an inverse verb in which the usual *ŋga-* is replaced by the third person ergative prefix *mbu-* or its variants when the object is a local person (see section 5.2.2.3); this example also dispenses with the perfective prefix *t-*, functioning simply with the suffixal component *–andə*, a not uncommon feature in the narrative texts. In both (5.29e, f), as they are inverse, the person and number of the object is indicated, and as expected with nominative forms, specifically by the suffixal component of the nominative marking for perfective verbs: Ø 1SG and *-nde* 1PL.

There are also a few other examples of third person ergative marking for inverse transitive verb forms in the perfective in the corpus. They involve a lower ranked subject, second or third person, acting on a first person object. Note the inverse forms of (5.30a, b) use the alternative for third person subjects in combination with local objects exemplified in (5.18), although in (5.30a) the subject is actually second person singular, so this example like (5.30c) involves impersonalization to third person. In both cases, the suffixal agreement is for the first person paucal object. (5.30c) is a very interesting example of a ditransitive verb in the perfective in which the recipient, although case marked obliquely by the dative postposition ŋga, is also indicated by the first singular allomorph of the perfective suffix -a; there is otherwise no suffixal marking for first person singular.

(5.30) (a) *inda aratəkəmb təmbitenandəkəkə*
inda aratək-mb t-mbi-t-e-n-andə-kəkə
house good-OBL PFV-3.ERG=2/1.OBJ-CAUS-come-LK-PFV-1PC
'you (SG) have brought us (PC) to a good house'

(b) *mana nana mbaenandəkək*
mana nana mba-e-n-andə-kəkə
ma-na nana 3.ERG/1.OBJ-come-LK-PFV-1PC
'my mama has come to us (PC)'

(c) *inda aratək mana ŋga*
inda aratək ma-na ŋga
house good 1SG-POSS DAT
tumbuturikera
t-mbu-t-riker-a
PFV-3.ERG=2/1.OBJ-CAUS-get.up-PFV
'you (SG) have erected for me a good house'

5.2.4 The dative suffixes

The dative suffixes were never obtained in elicitation sessions and only found on analyzing the narrative texts, but there they are pervasive and serve a number of functions. Because their only occurrence in the corpus is in the legends of the narrative texts where they express predominantly non-local persons, with local persons only occurring in sporadic direct quotes, the full paradigm was not obtained. But the forms that do occur in the corpus are set out in Table 9:

5.2 The pronominal affix agreement systems for core arguments — 91

Table 9: Dative Suffixes.

	SG	DL	PC	PL
1	-ananga			
2	-anana			
3	-ana	-anəmba	-anəngra	-anumbwa

The third dual suffix has an alternative form *-anda*. The initial /a/ of these suffixes is always lost following a [-high] vowel.

These dative suffixes are pronominal agreement affixes used to indicate a human participant that is affected by the action or toward whom the action can be said to be directed. In view of this semantics, one might believe that they would be better labeled as object pronominal agreement suffixes, but their actual usage does not support that claim. The main reason is that the presence of a dative suffix does not convert an intransitive verb into a transitive one. Consider the following example:

(5.31) akən   asawa   upratanəmbakoya
        akən   asawa   u-pra-ta-anəmba-k-oya
        sun   close   3SG-excrete-OUT-3DL.DAT-FR.PAST-3SG
        'the day was about to dawn on them (DL)'

The verb *pra-ta-* excrete-OUT 'to dawn' is intransitive and is inflected as such in example (5.31) with the prefix *u-* 3SG (see Table 4); a transitive verb with a third person singular subject would occur with the prefix *mbu-* 3.ERG (see Table 6). The dative suffix in (5.31) does not affect the transitivity of this verb and hence should not be analyzed as indicating agreement for an object grammatical function. Similarly, a verb like *sara-* 'to report, narrate' can inflect intransitively or transitively, but the presence of a dative suffix does not require transitive inflection:

(5.32) (a) mina   ŋga   enamb   mbusaranaŋgənaya
             mi-na   ŋga   ena-mb   **mbu**-sara-n-aŋg-naya
             2SG-POSS   DAT   PROX.SG-OBL   3.ERG=1-report-LK-PRES-2SG
             'I tell you (SG) about this (SG)'

       (b) mu   usaranəmba-k
             mu   **u**-sara-anəmba-k
             3SG   3SG-report-3DL.DAT-FR.PAST
             'he told them (DL)'

In (5.32a) *sara-* 'report, narrate' is inflected transitively with a first person singular subject acting on a second person object, here the recipient of the reported speech. This, of course, is a configuration triggering impersonalization (see section 5.2.2.5), so the subject agreement is *mbu-* 3.ERG. In (5.32b), the recipient of the reported speech is third person dual and indicated by the dative suffix *-anəmba*. Note that the verb is inflected intransitively with *u-* 3SG, demonstrating again that these dative suffixes do not affect transitivity and hence should not be analyzed as pronominal object agreement markers. A related issue is the obvious formative *-an* that initiates all of these dative suffixes; should this be analyzed as an applicative marker to which is then appended the pronominal agreement suffix, i. e. *-anəmba* < *-an* APPL + *-mba* 3DL? While this may be a plausible historical scenario, it cannot be correct synchronically, as both (5.31) and (5.32b) demonstrate that these dative suffixes do not increase the valence of a verb.

The dative suffixes have a wide range of functions, largely parallel to that of dative cases in other languages. However, the dative suffixes for the local persons in the corpus are mainly found in their use in possessor raising; they do not appear to be able to mark recipients or benefactives. When local persons occur as recipients or benefactives, direct or impersonal inflections of transitive verbs are required. Dative suffixes mark non-local recipients of ditransitive verbs like *kam-* 'give':

(5.33) (a) nda itəmandək enana mbukamanak
nda itəmandək enana mbu-kam-ana-k
and younger.sister ♀ PROX.SG 3.ERG-give-3SG.DAT-FR.PAST
'and (her) younger sister gave this to her'

(b) iniŋ ŋgikamanak Wak
iniŋ ŋgi-kam-ana-k Wak
stone 3PL-give-3SG.DAT-FR.PAST PN
'they (PL) gave the stone to Wak'

(c) mayn yowa ŋgipukoranandə
mayn yowa ŋgi-pukora-ana-andə
male DIST 3PL-give.food-3DAT-SEQ
'they (PL) gave food to that man and then'

They also indicate other types of non-local recipients such as the target of reported speech or a person toward whom motion (or emotion) is directed:

(5.34) (a) enamb   usaranəŋgrakoya
           ena-mb   u-sara-nəŋgra-k-oya
           PROX.SG-OBL   3SG-report-3PC.DAT-FR.PAST
           'he told them (PC) about this'

   (b) usiranəmbak
       u-siri-anəmba-k
       3SG-go.down-3DL.DAT-FR.PAST
       'he came down to them (DL)'

   (c) musitkayn   urikeranəmbak
       musit-kayn   u-riker-anəmba-k
       dolphin-canoe   3SG-get.up-3DL.DAT-FR.PAST
       'the dolphin canoe came up to them (DL)'

   (d) urəmanumbwak   kar   erakot   ŋga
       u-rəmə-anumbwa-k   kar   e-ra-kot   ŋga
       3SG-stand-3PL.DAT-FR.PAST feast   come-stay-carry   DAT
       'he stood up before them (PL) in order to bring the feast'

   (e) Matawr   ukararanumbwak
       Matawr   u-karara-anumbwa-k
       PN.clan 3SG-ask-3PL.DAT-FR.PAST
       'he asked them (PL) of the Matawr clan'

   (f) ukusamenəmbakoya
       u-kusa-am-e-anəmba-k-oya
       3SG-go.out-DETR-come-3DL.DAT-FR.PAST-3SG
       'he came out and came towards them (DL)'

   (g) ambisen   mu   mbutumuranak
       ambisen   mu   mbu-t-mur-ana-k
       daughter 3SG 3.ERG-COM-fear-3SG.DAT-FR.PAST
       'the daughter took fright of that'

   (h) munəmb   yowa   ukusamanəmbakoya
       munəmb   yowa   u-kusa-am-anəmba-k-oya
       smell   DIST   3SG-go.out-DETR-3DL.DAT-FR.PAST
       'that smell went over to them (DL)'

They also express the beneficiaries of an action:

(5.35) (a) *indan mbuturikeranak*
indan mbu-t-riker-ana-k
house 3.ERG-CAUS-get.up-3SG.DAT-FR.PAST
'she erected a house for her'

(b) *nana nəmbre okumbi mbutetanak*
nana nəmbre okumbi mbu-t-e-ta-ana-k
mama pig PL 3.ERG-CAUS-come-OUT-SG.DAT-FR.PAST
'mother brought pigs out for him'

(c) *mumoran mbutadabanəŋgrak*
mə-moran mbu-tandamba-anəŋgra-k
eat-thing 3.ERG-prepare-3PC.DAT-FR.PAST
'he prepared all the food for them (PC)'

(d) *porakayn upwakamanəmbak*
porakayn u-pwak-am-anəmba-k
path 3SG-open-DETR-3DL.DAT-FR.PAST
'the path became clear for them (DL)'

(e) *tarar mbukaratanumbwak*
tarar mbu-karat-anumbwa-k
stone.axe 3.ERG-sharpen-3PL.DAT-FR.PAST
'he sharpened the stone axe for them (PL)'

(f) *inamasen e usianəmbwak*
inamasen e u-si-anəmbwa-k
knife LIKE 3SG-become-3PL.DAT-FR.PAST
'it became like a knife for them (PL)'

The dative suffixes also indicate human particpants, local (5.36f) or non-local, to which an uncontrolled event happens. This is most common with weather or time verbs (5.36a, b):

(5.36) (a) *akən upratanəmbak*
akən u-pra-ta-anəmba-k
sun 3SG-excrete-OUT-3DL.DAT-FR.PAST
'the day dawned upon them (DL)'

5.2 The pronominal affix agreement systems for core arguments — 95

(b) *akən usisenumbwak*
akən u-si-se-anumbwa-k
sun 3SG-become-NIGHT-3PL.DAT-FR.PAST
'night fell upon them (PL)'

(c) *miɲjir pra ŋga usianakoya*
miɲjir pra ŋga u-si-ana-k-oya
urine excrete DAT 3SG-do-3SG.DAT-FR.PAST-3SG
'she felt like urinating'

(d) *kanoŋ yowa urotanakoya*
kanoŋ yowa u-rot-ana-k-oya
shell DIST 3SG-fall-3SG.DAT-FR.PAST-3SG
'that shell fell down from him'

(e) *awr upretanəmbak*
awr u-pre-t-anəmba-k
fire 3SG-die-APPL-3DL.DAT-FR.PAST
'the fire died on them (DL)'

(f) *kandəknambrin usinaŋgak*
kandək-nambrin u-si-anaŋga-k
sleep-eye 3SG-do-1SG.DAT-FR.PAST
'I was sleepy' (literally 'sleep-eye did to me')

Finally, these dative suffixes are used frequently in possessor raising constuctions in which the human possessor, most commonly of a body part but not restricted to these (5.37c, d, e, f), is indicated by agreement with a dative suffix (see section 5.5.2 on possessor raising). These constructions permit local participants to be indicated by a dative suffix (5.37c):

(5.37) (a) *namən mbwerakotanak*
namən mbu-e-ra-kot-ana-k
leg 3.ERG-come-stay-carry-3SG.DAT-FR.PAST
'she brought her leg'

(b) *kambowen asumb o ndək ŋgiparianak*
kambowen asumb o ndək ŋgi-pari-ana-k
spines mouth PURP 3PL-extract-3SG.DAT-FR.PAST
'they (PL) removed the spines from his mouth'

(c) ara   ŋga   apən   isambotmananaomena
    ara   ŋga   apən   i-sambo-t-ma-anana-ome-ona
    what   DAT   food.bits 2-leave-APPL-DUR-2SG.DAT-*wh*-2SG
    'why do you (SG) keep leaving food bits of yours (SG)?'

(d) irormandək   usiranak   ŋaɲjen
    irormandək   u-siri-ana-k   ŋaɲjen
    tree.spirit 3SG-go.down-3SG.DAT-FR.PAST child
    mbwerakotanak
    mbu-e-ra-kot-ana-k
    3.ERG-come-stay-carry-3SG.DAT-FR.PAST
    'the tree spirit came down to her and brought her child'

This could be a recipient dative rather than possessor raising ('brought the child to her')

(e) nimbep   napar   sur   o   ndək
    nimbep   napar   sur   o   ndək
    spear   hand   inside   PURP
    urotanakəto
    u-rot-ana-k-to
    3SG-fall-3SG.DAT-FR.PAST-DEP
    'the spear fell from his hand and'

(f) uren   mbubagarianakondu
    uren   mbu-mbaŋgari-ana-k-ondu
    dog   3.ERG-kill-3SG.DAT-FR.PAST-3PL
    'they (PL) killed her dog'

The last example could be malefactive ('killed the dog on her'), rather than strictly possessor raising.

5.3 Tense, aspect and mood

Kopar has a quite rich system of tense, aspect and mood distinctions, and this is certainly one area of the grammar where much more extensive fieldwork would be needed to gain a firmer understanding. Aspect in particular is only partially understood, so I will focus on those aspectual contrasts which are more straightforward. I will separate the category of mood here into modality, which expresses the reality status of events, which on the basis of the current data, essentially corresponds to obligation, ability, permission and negation, and mood or illocutionary force, the nature of the speech act being performed, statement, question, command, etc.

5.3.1 Tense

Other than with the negative verbs *kay-* or *nda-*, tense is generally an obligatory inflection for finite verbs in Kopar. All tense marking is suffixal, in the penultimate suffix slot before a bound pronominal suffix or dependent verb suffixes, if present. If these are lacking, tense is in word final position (5.38b):

(5.38) (a) *mbusamaytəndəkənaya*
mbu-samayt-ndək-naya
3.ERG=1-see-NR.PAST-2SG
'I/we saw you (SG)'

(b) *ŋgiesek*
ŋgi-ese-k
3PL-bring-FR.PAST
'they (PL) brought it'

There are six tenses in Kopar: far past, near past, present, perfective, immediate future and future. As with Yimas (Foley 1991:236–238), strictly speaking tense is a subset of a modality contrast in Kopar between unreal and real events, a contrast that is recast in Kopar in the distribution of the bound pronominals. As we saw in section 5.2, bound pronominals are split for their case alignment, ergative versus accusative, according to tense-aspect-mood type, ergative for realized tenses and aspects and accusative for unrealized tenses and moods. For tense marking, though, the contrast is between marking events which are seen as unreal, not bound in the experienced temporal continuum, and those which are real, within the experienced temporal continuum. This is clearly seen by the distribution of the suffix *-k*, which is used for events in a far off, not witnessed, time, very common in traditional myths and legends (5.39a) (glossed far past FR.PAST), for hypothetical or indefinite events in a possible future time (5.39b) (glossed irrealis IRR), and for adjectival verbs which denote properties, i.e. states not bound in time (5.39c) (also glossed irrealis):

(5.39) (a) nəmbən   budukondu
           nəmbən   mbu-ndə-k-ondu
           garamut.drum 3.ERG-hear-FR.PAST-3PL
           'they heard the garamut drum'

   (b) sokay   məkəmb   o   mak   simbina
       sokay   mə-k-mb   o   mak   si-mbi-ona
       tobacco eat-IRR-OBL bad become-IM.FUT-2SG
       'by smoking tobacco, you will get ill' (literally 'become, bad')

   (c) naŋgun mana   kaymbakoya
       naŋgun ma-na   kaymba-k-oya
       skin   1SG-POSS   white-IRR-3SG
       'my skin is white'

In summary, the suffix -*k* prototypically marks events whose temporal coordinates may not be fixed by the speaker in experiential time, most notably in the further not witnessed or less vividly remembered past and the more indefinite less probable future. Because this occurs at both ends of the temporal continuum, as with Yimas, we may conjecture that Kopar speakers view time not as an open infinitely expandable line, but as a closed circle, as in Figure 4:

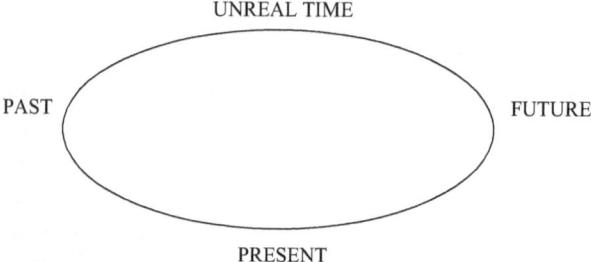

Figure 4: The temporal continuum in Kopar.

From the standpoint of 'now' in the present, Kopar speakers look back into the past toward an unknown mythical, yet realized, time and into the future toward another unknown hypothetical, and unrealized, time, but both of these ultimately converge on events that are fundamentally viewed as unreal, not within experiential time, but constructed on the basis of received transmission, conjecture or potentially faulty memory.

5.3.1.1 The far past tense
This tense is marked by the suffix -*k* and is used for events from the day before yesterday all the way to those in mythical or legendary times of the ancestors. It is most commonly found in narratives depicting myths and legends, in fact all four narrative texts were given in this tense, but can refer to events as recently as the day before yesterday. As a tense expressing realized events, it always occurs with bound pronominals in an ergative alignment:

(5.40) (a) *mu utandək upipikambusekoya*
mu uta-ndək u-pipikambu-se-k-oya
3SG 2.days.removed-PAST 3SG-fish.with.hook-NIGHT-FR.PAST-3SG
'he fished at night the day before yesterday'

(b) *məŋgə mbusamaytukondu andimeŋ sur*
məŋgə mbu-samayt-k-ondu andimeŋ sur
3PC 3.ERG-see-FR.PAST-3PL male.cult.house inside
'They (PL) saw them (PC) inside the male cult house'

5.3.1.2 The near past tense
The near past is used for events which took place yesterday and is indicated by the suffix -*ndək*, subject to the vowel harmony rules discussed in section 2.4. This seems to be the unmarked past tense marker as indicated by its extension to temporal words to mark past time (section 3.8), even to *uta* '2 days removed' (example (5.40a) above) to indicate previous days. Again, as a tense for realized events, it always occurs with bound pronominals in an ergative alignmment:

(5.41) (a) *tamənd irumondindukoke*
tamənd i-ru-mondi-ndək-oke
fish 1-shoot-penetrate-NR.PAST-1DL
'we (DL) speared fish yesterday'

(b) *ma mi rarindək mbusamaytəndəkənaya*
ma mi rari-ndək mbu-samayt-ndək-naya
1SG 2SG 1.day.removed-PAST 3.ERG=1-see-NR.PAST-2SG
'I saw you yesterday'

5.3.1.3 The present tense
The present tense is used for events happening today or habitual events or true generic events and is marked by the suffix -*aŋ*. Again, as one of the tenses that expresses realized events, it occurs with bound pronominals inflecting on an

ergative alignment. Following a verb root or suffix that ends in a non-high vowel, /n/ insertion (rule 2.9) applies. When used to express events happening right now, the present tense suffix commonly co-occurs with the progressive aspectual suffix -*ar*, though that is not obligatory. Consider the following contrasting examples, without -*ar* PROG (5.42a) and with it (5.42b):

(5.42) (a) nor   aynde   miɲjir   upranaŋgoya
           nor   aynde   miɲjir   u-pra-n-aŋg-oya
           man   here   urine   3SG-excrete-LK-PRES-3SG
           'the man urinates/is urinating here'

      (b) nor   aynde   miɲjir   upraraŋgoya
           nor   aynde   miɲjir   u-pra-ar-aŋg-oya
           man   here   urine   3SG-excrete-PROG-PRES-3SG
           'the man is urinating here (right now as we speak)'

Example (5.42a) can either have a habitual reading or an ongoing progressive reading, though the former is favored here, while (5.42b) with an overt progressive suffix -*ar* can only have the latter reading. Here is another example (5.43a) of the favored habitual use of the present tense and one of a true generic usage:

(5.43) (a) nor   nəmbren   mbumaŋgoya
           nor   nəmbren   mbu-mə-aŋg-oya
           man   pig   3.ERG-eat-PRES-3
           'the man eats pork/ is eating a pig'

      (b) taimənd   masak   sur   utapwataŋgoya
           tamənd   masak   sur   u-tapwat-aŋg-oya
           fish   ocean   inside   3SG-swim-PRES-3SG
           'fish swim in the ocean'

When the present tense is used to express an event happening today, it has only a very narrow window around the temporal point of 'now'. If the event is already finished, then instead the perfective (section 5.3.1.4) is required, and if it has not yet occurred, but should or is widely expected to occur very soon today, then the immediate future (section 5.3.1.5) would be employed. Further, when the present tense suffix occurs word finally, i. e. there is no following bound pronominal, commonly it is truncated to simply -*a*:

(5.44) (a) *nor   undasara*
         nor   u-ndasa-ar-a
         man   3SG-sit-PROG-PRES
         'the man is sitting'

   (b) *mu   ona*
       mu   o-n-a
       3SG go-LK-PRES
       'he is going'

   (c) *nəmandək   nor   budara*
       nəmandək   nor   mbu-ndə-ar-a
       woman   man   3.ERG-hear-PROG-PRES
       'the woman is listening to the man'

5.3.1.4 The perfective aspect

While the perfective is semantically an aspect, it is treated here because formally it behaves like a tense in Kopar. The perfective aspect is used for events which have been recently completed, particularly today. Because of its semantics of recently completed events, it is commonly used to denote accomplishments, events which express a transition from one state or activity to another (Dowty 1979; Foley and Van Valin 1984; Vendler 1967). The perfective belongs to the set of tenses for realized events, and so bound pronominals with it inflect on an ergative basis, but with a somewhat different pattern than the other realized tenses (see section 5.2.3). It is marked by a circumfix, a prefixal component *t-* and a suffixal component *-andə ~ -a* (*-a* with first singular and third person, *-andə* elsewhere), which triggers nasal insertion (rule 2.9) following non-high vowels, though one or the other of these affixes can be omitted. The suffixal component also functions to mark sequence for dependent verbs in clause chaining coordinate constructions (see section 7.2.3).

(5.45) (a) *nəmandək   təndasana*
         nəmandək   t-Ø-ndasa-n-a
         woman   PFV-3SG-sit-LK-PFV
         'the woman sat down'

   (b) *təmakandəka*
       t-ma-kandək-a
       PFV-1SG-sleep-PFV
       'I fell asleep'

(c) *kayn təpakanandanana*
kayn t-paka-n-andə-anana
canoe PFV-carve-LK-PFV-2SG
'you (SG) have carved a canoe'

(d) *indan tundiandikike*
indan t-undi-andə-kike
house PFV-build-PFV-1DL
'we (DL) have built a house'

(e) *təmao*
t-ma-o
PFV-1SG-go
'I have gone'

(f) *asawa təsiandəma*
asawa t-si-andə-ma
close PFV-do-PFV-1SG
'I've almost done it'

(g) *mu təprek*
mu t-Ø-pre-k
3SG PFV-3SG-die-FR.PAST
'he/she had already died'

Note in (5.45g) there is a combination of both the perfective aspect prefix and the far past tense, to indicate a completed event in far past time, equivalent to a pluperfect tense in languages like Latin.

5.3.1.5 The immediate future tense

The immediate future is used for events which are expected to occur very soon, most commonly in a closely following time period of today, but not restricted to today (see 5.46d). Because of this strong expectation, it often has a semi-modal force, best translated by 'should'. As one of the unrealized tenses, it always occurs with bound pronominals which are accusatively aligned. It is indicated by the suffix *-mbi* (*-mba* in the Wongan dialect), subject to denasalization (rule 2.12) when a following bound pronominal begins with a voiced prenasalized stop (5.46b-d):

(5.46) (a) *ma ombiya*
ma o-mbi-oya
1SG go-IM.FUT-1SG
'I should go/I'm going to go'

(b) *mora mə-bi-də*
mora mə-mbi-ondə
thing eat-IM.FUT-3DL
'they (DL) should eat/ are going to eat'

(c) *ruaŋ saytəbidu*
ruaŋ sayt-mbi-ondu
coconut pick-IM.FUT-3PL
'they (PL) should pick/ are going to pick a coconut'

(d) *rari ebida*
rari e-mbi-onda
1.day.removed come-IM.FUT-3SG
'he should come/is going to come tomorrow'

Because of the semi-modal force semantics of the immediate future, i. e. 'should', it is often used with non-singular second person subjects as a less bald imperative (see section 5.3.3):

(5.47) (a) *ŋgu mobiduku*
ŋgu ma-o-mbi-onduku
2PC ITR.IMP-go-IM.FUT-2PL
'you (PC) should go!'

(b) *ruaŋ kusaytəmbiko*
ruaŋ ku-sayt-mbi-oko
coconut TR.IMP-pick-IM.FUT-2DL
'you (DL) should pick a coconut!'

5.3.1.6 The future tense

The future tense is for any event that will occur in future time, spanning from tomorrow to an indefinite time in the future. It contrasts with irrealis *-k* in that the event is more than merely hypothetical. There are reasons to believe that it will occur within the continuum of experiential time, but it is not strongly expected in the very near future as with the immediate future *-mbi*. As it is a tense for unrealized events, it too occurs with bound pronominals having an accusative alignment. The marker for the future is *-onduk*, made up of *-o*, plausibly originally the verb *o-* 'go' (see the use of 'go' to indicate the future in both English ('be going to') and colloquial French (*je vais manger* 'I'm going to eat') and *-nduk*, via vowel harmony from *-ndək*, homophonous with the near past, the tense for events occurring yesterday. But this homophony would not be surprising given *rari* 'one day removed', denoting both 'yesterday' and 'tomorrow'. However, it is perhaps more likely that *-ndək* actually derives from the purposive postposition *ndək*. Some examples of the future are:

(5.48) (a) uri   nambratondukonde   rari
            uri   nambrat-onduk-onde   rari
            crocodile   spear-FUT-1PL   1.day.removed
            'we (PL) will spear the crocodile tomorrow'

      (b) mi   indan   undiodukona
           mi   indan   undi-onduk-ona
           2SG   house   build-FUT-2SG
           'you (SG) will build a house'

      (c) rari   akən   mana   naŋgun   pet
           rari   akən   ma-na   naŋgun   pet
           1.day.removed   sun   1SG-POSS   skin   dark/black
           təkamondukonda
           t-kam-onduk-onda
           CAUS-become-FUT-3SG
           'the sun will darken my skin tomorrow'

5.3.2 Aspect

From the data available, aspect unfortunately still remains not fully understood. Only four aspectual distinctions, the perfective discussed in section 5.3.1.4 and the progressive, durative and extended aspects described below, are at this point well established. It is quite likely that as a moderately polysynthetic language with incorporation, other aspectual categories in Kopar are expressed through bound adverbials, as in Yimas or Alamblak (Bruce 1984); some of these have been identified and will be described in section 5.4.2.

5.3.2.1 Progressive aspect

Progressive aspect is indicated by a suffix *-ar*, which immediately follows the verb stem and typically loses the initial /a/ if the verb stem ends in /a/. The progressive aspect indicates an event is in the process of dynamically unfolding and has not yet reached an end point or state. Unlike the perfective aspect which is normally restricted to events recently completed today, though there are rare exceptions (5.45g), the progressive aspect can occur with any tense, though far and away most commonly with the present tense.

(5.49) (a) məŋgə indan ŋgiundiaraŋoya
           məŋgə indan ŋgi-undi-ar-aŋg-oya
           3PC   house 3PC-build-PROG-PRES-3
           'they (PC) are building a house'

   (b) ma   ŋgu   mbusamaytarəndikiya
       ma   ŋgu   mbu-samayt-ar-ndək-iya
       1SG 2PC   3.ERG=1-see-PROG-NR.PAST-PC
       'I was watching you (PC) yesterday'

   (c) ma   mora   marəmbiya   rari
       ma   mora   mə-ar-mbi-oya   rari
       1SG thing eat-PROG-IM.FUT-1SG   1.day.removed
       'I should eat/am going to eat tomorrow'

   (d) ara   mbusiaraŋomendu
       ara   mbu-si-ar-aŋg-ome-ondu
       what   3.ERG-do-PROG-PRES-*wh*-3PL
       'what are they (PL) doing?'

5.3.2.2 Durative aspect

In addition to *-ar* PROG, there is another aspectual suffix which indicates an imperfective type of aspect, which may best be described as durative aspect and in most contexts is best translated as 'continue to' or 'keep on'. Its form is *-ma* DUR. Here are a few examples:

(5.50) (a) indan   naundimanaŋoya
           indan   na-undi-ma-n-aŋg-oya
           house   1SG.ERG-build-DUR-LK-PRES-1SG
           'I'm continuing to build a house'

   (b) ma   utu   porakay   makaromakənaya
       ma   utu   porakay   ma-kar-o-ma-k-naya
       1SG 2.days.removed   path   1SG-walk-go-DUR-FR.PAST-1SG
       'I was walking along the path two days ago'

(c) mumoran   kutumǝndǝmana
    mǝ-moran   ku-t-mǝnd-ma-ona
    eat-thing   TR.IMP-CAUS-finish-DUR-2SG
    'you (SG) must continue to finish the food!'

(d) ŋgu   paomanaŋgi
    ŋgu   pa-o-ma-n-aŋg-iya
    2PL   STILL-go-DUR-LK-PRES-PC
    'you (PC) still keep on going'

(e) ko   arǝm   inamanaŋgoko
    ko   arǝm   ina-ma-n-aŋg-oko
    2DL   water   fill-DUR-LK-PRES-2DL
    'you (DL) keep filling up the water'

(f) uramak
    u-ra-ma-k
    3SG-stay-DUR-FR.PAST
    'it remained'

(g) ambisen   ukandǝksemakoya
    ambisen   u-kandǝk-se-ma-k-oya
    daughter   3SG-sleep-NIGHT-DUR-FR.PAST-3SG
    'the daughter kept on sleeping during the night'

5.3.2.3 Extended aspect

This aspect typically occurs in combination with *-ar* PROG, and indicates that the dynamic process of the event or action unfolding continues for an extended period of time. It is marked by a reduplicated from of the verb root *o-* 'go' separated by an intrusive /r/, hence *-oro* EXT. It corresponds closely in meaning to the similar use of reduplicated *go* in Tok Pisin, i. e. *i go go go*, and may very well ultimately be a calque on that Tok Pisin structure. Here are some examples:

(5.51) (a) urarorok
           u-ra-ar-oro-k
           3SG-stay-PROG-EXT-FR.PAST
           'he was staying for a while'

       (b) mǝŋgǝ   ŋgipunuarorok
           mǝŋgǝ   ŋgi-punu-ar-oro-k
           3PC   3PC-work.sago-PROG-EXT-FR.PAST
           'they (PC) were working sago for a while'

(c) *Wak yo mbutukararorokondu*
Wak yo mbu-t-kar-ar-oro-k-ondu
PN DEF 3.ERG-COM-walk-PROG-EXT-FR.PAST-3PL
'they (PL) were walking around with Wak for a while'

(d) *kiŋgep mbuturaporaposarorokondu*
kiŋgep mbu-t-rapo+rapo-sa-ar-oro-k-ondu
ladder 3.ERG-COM-run.RED-IN-PROG-EXT-FR.PAST-3PL
'they were running in with a ladder for a while'

(e) *mu naprəkarorokoya*
mu na-prək-ar-oro-k-oya
3SG 3SG-work-PROG-EXT-FR.PAST-3SG
'he was working (on something) for a while'

(f) *gibabasarorok*
ŋgi-mbambas-ar-oro-k
3PL-confused-PROG-EXT-FR.PAST
'they were confused for a while'

5.3.3 Modality

Modality describes the actuality of an event, whether it is realized or not. We have already seen in section 5.2 how important this notion is in determining the inflection of bound pronominals in Kopar. In a semantic sense, both the immediate future and future tense are modal categories, as they expressed unrealized events, but in Kopar these pattern formally with the other realized tenses in occupying the same suffixal slots, so have been treated under tense in section 5.3.1. Modality is generally concerned with concepts that describe the nature of unrealized events: are they necessary, likely, possible or have not actually occurred at all? The last of these is, of course, negation.

5.3.3.1 Negation
Negation is expressed in Kopar with the negative verb *kay-* NEG, typically taking the clause to be negated as its complement. As *kay-* NEG is a verb for unrealized events, it takes the set of bound pronominals for unrealized events of Table 3:

(5.52) (a) mu   rari   arəm   kirionduk   kaynda
           mu   rari   arəm   ki-ri-onduk   kay-onda
           3SG   1.day.removed   water   bathe-DOWN-FUT   NEG-3SG
           'he won't bathe tomorrow'

   (b) ma   ndə   kaya
       ma   ndə   kay-oya
       1SG   hear   NEG-1SG
       'I didn't listen'

   (c) rarindək   arəm   kiri   kaynda
       rari-ndək   arəm   ki-ri   kay-onda
       1.day.removed-NR.PAST   water   bathe-DOWN   NEG-3SG
       'he didn't bathe yesterday'

   (d) rari   onduk   kaynda   pipikambu   ŋga
       rari   o-onduk   kay-onda   pipikambu   ŋga
       1.day.removed   go-FUT   NEG-3SG   fish.with.hook   DAT
       'he/she won't go fishing tomorrow'

   (e) ke   ɲja   iramanaŋəbake   kayna
       ke   ɲja   i-ra-ma-n-aŋg-mbake   kay-ona
       1DL   just   1-stay-DUR-LK-PRES-1DL   NEG-2SG
       'just us (DL) continue to stay, you (SG) not'

   (f) karo   kaynda   imbotma
       kar-o   kay-onda   imbotma
       walk-go   NEG-3SG   jealousy
       usiaroronakoya
       u-si-ar-oro-ana-k-oya
       3SG-do-PROG-EXT-3SG.DAT-FR.PAST-3SG
       'he didn't walk around, he continually felt jealous' (literally 'jealousy did to him')

Note that in the negative, bound pronominal affixes are not realized on their governing verb but 'climb' to appear on the negator. The examples in (5.52) are the normal order with the negator following the main verb; however, it is possible for it to precede the verb, but even in this case, the bound pronominals appear on the negator. Compare (5.52b) with (5.53):

(5.53) (a) *ma   kaya   ndə*
           ma   kay-oya   ndə
           1SG   NEG-1SG   hear
           'I didn't listen'

      (b) *saran   kaynda   ndə*
          sara-n   kay-onda   ndə
          report-NMLZ   NEG-3SG   hear
          'he didn't listen to the report'

      (c) *saran   ombe   kaynda   kata   muna*
          sara-n   ombe   kay-onda   kata   mu-na
          talk-NMLZ   INDEF   NEG-3SG   speak   3SG-POSS
          'he did not say any talk of his' (Tok Pisin *em i nogat wanpela tok*)

5.3.3.2 Ability

Ability refers to the means of possible events becoming real. This concept is expressed similarly to negation, but this time by a noun *tayn* 'ability', derived from the verb *tay-* 'try' to which the nominalization suffix *-n* is attached. This noun is then suffixed with bound pronominals from the unrealized set of Table 3 to produce a verb meaning 'to have ability' (see section 6.2.4 on non-verbal possession), which takes the enabled action as a complement. However, the complement can either be left unmarked (5.54a), as with *kay-* NEG, or marked with the dative or purposive postpositions, as in the desiderative constructions of section 7.1.1–7.1.2 (5.54b, c, e).

(5.54) (a) *mina   nda   pipikambu   taynoya*
           mi-na   nda   pipikambu   tayn-oya
           2SG-POSS   COM   fish.with.hook   ABIL-1SG
           'I can go fishing with you (SG)'

      (b) *mu   taynonda   indan   undi   ŋga*
          mu   tayn-onda   indan   undi   ŋga
          3SG   ABIL-3SG   house   build   DAT
          'he can build a house'

      (c) *sina   mumora   nambri   si   ŋga   tayn   kay-ndu*
          sina   mə-mora   nambri   si   ŋga   tayn   kay-ondu
          daytime   eat-thing   eye   do   DAT   ABIL   NEG-3PL
          'they (PL) can't find food in the daytime'

(d) o   ŋga   tayn   kaya
    o   ŋga   tayn   kay-oya
    go   DAT   ABIL   NEG-1SG
    'I can't go'

(e) ma   mina   nda   ra   ndək   tayn   kaya
    ma   mi-na   nda   ra   ndək   tayn   kay-oya
    1SG   2SG-POSS   COM   stay PURP   ABIL   NEG-1SG
    'I can't stay with you (SG)'

Note that *tayn* 'ability' is overtly a noun in (5.54c, d, e), in which it is negated by *kay-* NEG, 'I don't have the ability'. The idiom *nambri(n) si-* eye do/make means 'to scan for, visually search for, find'.

5.3.3.3 Necessity or obligation

Necessity refers to the requirement that an event become real and there is an obligation or requirement on the subject to make the event become real. There is no one simple way to express this in Kopar; there are number of strategies to convey it available to speakers. Perhaps the most common is the immediate future. With its semi-modal entailment that the event is strongly expected to occur very shortly, an implicature of 'should' or even 'must' commonly holds:

(5.55) (a) *paŋgə mimbikə*
           paŋgə mə-mbi-okə
           1PC   eat-IM.FUT-1PC
           'we (PC) should eat'

       (b) *pipikambusembina*
           pipikambu-se-mbi-ona
           fish.with.hook-NIGHT-IM.FUT-2SG
           'you (SG) must fish tonight'

Another means to express necessity or obligation employs the incorporated adverbial *nda-* 'right then, now' used in combination with a suffix *-n ~ -na*, the latter allomorph similar to one of the allomorphs of the singular imperative suffix, but here not restricted to second person:

(5.56) (a) mu   mumoran   budaukina
           mu   mə-moran   mbu-nda-uki-na
           3SG   eat-thing   3.ERG-NOW-buy-IMP
           'he/she must buy food'

(b) ma   numot   sur   mandaon
    ma   numot   sur   ma-nda-o-n
    1SG village inside ITR.IMP-NOW-go-IMP
    'I must go to the village'

Finally, there are an emphatic preverbal particle *wa* NEC and a verbal suffix *-ma* that seem to have the same function:

(5.57) (a) mu   mumoran   wa   ukiondukoya   rari
           mu   mə-moran   wa   uki-onduk-oya   rari
           3SG eat-thing   NEC buy-FUT-3SG   1.day.removed
           'he must buy food tomorrow'

(b) mi   mumoran   wa   kukika
    mi   mə-moran   wa   ku-uki-ka
    2SG eat-thing   NEC TR.IMP-buy-SG.IMP
    'you (SG) must buy food!'

(c) makandəksenama
    ma-kandək-se-ona-ma
    ITR.IMP-sleep-NIGHT-2SG-NEC
    'you (SG) must sleep!'

(d) moran   kutumandəmana
    moran   ku-t-mənd-ma-ona
    food   TR.IMP-CAUS-finish-NEC-2SG
    'you (SG) must finish the food'

(e) sarapaki naŋgun uraŋgoyama
    sarapaki naŋgun u-ra-aŋg-oya-ma
    cold   skin   3SG-stay-PRES-3SG-NEC
    '(my) skin must be cold'

Note the variable placement of *-ma* NEC with respect to the pronominal agreement suffix in (5.57c, d).

5.3.3.4 Permissive modality
This expresses the modal meaning of 'be allowed to'. It only occurs a few times in the corpus, once in in a prohibitive (5.58b), but its form *ka-* PERM is cognate with the Yimas prefix of the same meaning (Foley 1991:265–268):

(5.58) (a) kandaen   kaysar   kandandasan
          ka-nda-e-n   kay-sar   ka-nda-ndasa-n
          PERM-NOW-come-IMP   canoe-rudder   PERM-NOW-sit-IMP
          ke   mandaŋgareke
          ke   ma-nda-ŋgara-e-ke
          1DL   ITR.IMP-NOW-front-come-1DL
          'you (SG) can come now, you (SG) can sit now at the rudder of the canoe, and let us (DL) come in front now'

    (b) kandawarəkəmbaya
        ka-nda-o-arək-mbaya
        PERM-NOW-go-PROHIB-DL
        'you (DL) can't go just now'

5.3.4 Mood or illocutionary force

Mood or illocutionary force refers to the type of speech act performed by an utterance, whether it is a statement, question, command or exhortation, among others. Statements have no overt segmental marking in Kopar; they represent the neutral speech act. They do, however, have a distinctive rising-falling pitch at the end of the utterance. There are four types of speech acts in Kopar which do have overt marking: questions, commands, prohibitions and exhortations.

5.3.4.1 Questions

Yes-no questions are indicated by a final suffix -e, which commonly replaces a final /a/ of a suffixal bound pronominal (5.59a):

(5.59) (a) mi   arəm   kiri   ŋga   rine?
           mi   arəm   ki-ri   ŋga   ri-ona-e?
           2SG   water   bathe-DOWN   DAT   DES-2SG-Q
           'do you (SG) want to bathe?'

    (b) məndə   ipraraŋgəbakwe?
        məndə   i-pra-ar-aŋg-mbako-e?
        feces   2-excrete-PROG-PRES-2DL-Q
        'are you (DL) defecating?'

(c) *təsiandədwe*
t-si-andə-ndo-e
PFV-do-PFV-2PL-Q
'have you (PL) already done it?'

(d) *mbiramanande*   e   *mbiprenande*
mbi-ra-ma-n-andə-e   e   mbi-pre-n-andə-e
3DL-stay-DUR-LK-SEQ-Q   or   3DL-die-LK-SEQ-Q
'are they (DL) still alive or have they (DL) died?'

(e) *ŋguna*   *ŋga*   *prerambiye*
ŋgu-na   ŋga   prera-mbi-oya-e
2PC-POSS   DAT   angry-IM.FUT-1SG-Q
'am I going to be angry with you (PC)?'

Examples of wh-questions in the corpus always occur with the *wh*-formative *-ome(na)* mentioned in section 3.5. It is suffixed either to a question word which functions like a verb root (5.60a) or to a verb like *si-* 'make, do, become, feel, happen' to mean 'do what, what happens' (5.60d, e, f, h) or a specific verb root (5.60b, c, g, i, j, k). Note that tense marking is sometimes omitted in wh-questions (5.60a, i, k, l):

(5.60) (a) *ndataromena*
ndata-ar-ome-ona
where-PROG-*wh*-2SG
'where are you (SG) going?' (literally 'you are where-ing?')

(b) *paŋgə ndata makisamek*
paŋgə ndata ma-kisa-ome-okə
1PC   where   ITR.IMP-hide-*wh*-1PC
'where should we (PC) hide?'

(c) *ndata makandəksemekə*
ndata ma-kandək-se-ome-okə
where ITR.IMP-sleep-NIGHT-*wh*-1PC
'where shall we (PC) sleep?'

(d) *aran mbusiaraŋgomendu*
aran mbu-si-ar-aŋg-ome-ondu
what   3.ERG-do-PROG-PRES-*wh*-3PL
'what are they (PL) doing?'

(e) mi   aramoran   isisenaŋgomena
    mi   ara-moran   i-si-se-n-aŋg-ome-ona
    2SG   what-thing   2-do-NIGHT-LK-PRES-*wh*-2SG
    'what are you (SG) doing?'

(f) ara-moran   usisenaŋgome
    ara-moran   u-si-se-n-aŋg-ome
    what-thing   3SG-happen-NIGHT-LK-PRES-*wh*
    'what's happening?'

(g) ke   nde   maomeke
    ke   nde   ma-o-ome-oke
    1DL   how   ITR.IMP-go-*wh*-1DL
    'how shall we (DL) go?'

(h) mi   dade   munde   isindəkoname
    mi   ndande   munde   i-si-ndək-ona-ome
    2SG   how.RED   there   2-do-NR.PAST-2SG-*wh*
    'how did you (SG) do (that) there?'

(i) nde   nakameme
    nde   na-kame-ome
    how   1SG.ERG-call.out-*wh*
    'how should I call (him) out?'

(j) ara   ŋga   imuraŋgime
    ara   ŋga   i-mur-aŋg-iya-ome
    what   DAT   2-afraid-PRES-PC-*wh*
    'why are you (PC) afraid?'

(k) ara   ŋga   apən   isambotmananamena
    ara   ŋga   apən   i-sambo-t-ma-anana-ome-ona
    what   DAT   food.bits   2-leave-APPL-DUR-2SG.DAT-*wh*-2SG
    'why do you (SG) keep leaving food bits of yours'

(l) meŋga   erakotime
    meŋga   e-ra-kot-iya-ome
    who.PC   come-stay-carry-PC-*wh*
    'who (PC) will bring (you)?'

Note the unusual placement of the *wh* element -*ome* after the pronominal agreement suffix in (5.60h, j, l)

5.3.4.2 Imperatives

Imperative utterances express commands, and their subjects are always second person. In Kopar imperative verbs are always marked for the number of their second person subjects by a set of bound pronominal suffixes:

Table 10: Bound Pronominal Suffixes for Imperative Mood.

SG	-ka or -ona
DL	-oko
PC:	-onduku
PL:	-ondo

The singular bound pronominal is sometimes omitted (5.61c). Note that, other than the singular form -ka, these pronominal suffixes are identical with the corresponding suffixes for the unrealized tenses in Table 3, and because these normally follow the immediate future suffix -mbi or sometimes a verb root or suffix ending in a vowel or semivowel (5.61f, k), the initial /o/ is elided. Verbs in imperative mood denote, of course, desired, but as yet unrealized events, so this is not unexpected. In addition, verbs in imperative mood, as generally in Kopar, distinguish between transitive and intransitive verbs. Intransitive verbs take a prefix ma-, while transitive verbs have a prefix ku-, though these too are sometimes omitted (5.61f). While plain imperatives with non-singular subjects are grammatical (5.61f), such commands, at least in the Kopar dialect, are much more commonly expressed in combination with the immediate future tense (5.61a, d, g, h, i). Very emphatic imperatives are expressed with the modal suffix for obligation or necessity -ma (examples (5.61k, l). Some examples are:

(5.61) (a) o   mobido   kumuobido
           o   ma-o-mbi-ondo   ku-mə-o-mbi-ondo
           2PL ITR.IMP-go-IM.FUT-2PL TR.IMP-eat-go-IM.FUT-2PL
           porakay yo
           porakay yo
           path   DEF
           'you (PL) go, eat while going along the path!'

       (b) maŋgaramaka
           ma-ŋga-ra-ma-ka
           ITR.IMP-FIRST-stay-DUR-2SG
           'you (SG) continue to stay first!'

(c) *ma kundə*
ma ku-ndə
1SG TR.IMP-hear
'listen to me!'

(d) *kutururujabiduku paret ŋari*
ku-t-ruruɲja-mbi-onduku paret ŋari
TR.IMP-CAUS-shake-IM.FUT-2PC outside DAT
'you (PC) shake (it) until (it comes) out!'

(e) *mapaŋgarikerəka*
ma-pa-ŋga-riker-ka
ITR.IMP-STILL-FIRST-get.up-2SG
'you get up right now!'

(f) *mana ŋga arəm ndateko*
ma-na ŋga arəm nda-t-e-oko
1SG-POSS DAT water NOW-CAUS-come-2DL
'you (DL) bring me water!'

(g) *sapikindi maŋgaebiduku*
sapiki-ndi ma-ŋga-e-mbi-onduku
good-ADV ITR.IMP-FIRST-come-IM.FUT-2PC
'it's good that you (PC) come right now!'

(h) *mambikisabiduku*
ma-mbi-kisa-mbi-onduku
ITR.IMP-AGAIN-hide-IM.FUT-2PC
'you (PC) go hide again!'

(i) *ko maŋgaombiko arəm*
ko ma-ŋga-o-mbi-oko arəm
2DL ITR.IMP-FIRST-go-IM.FUT-2DL water
maŋgakirimbiko
ma-ŋga-ki-ri-mbi-oko
ITR.IMP-FIRST-bathe-DOWN-IM.FUT-2DL
'you (DL) go first now and bathe!'

(j) *porakay yo tak kureramoka*
porakay yo tak ku-rer-am-o-ka
path DEF betel.refuse TR.IMP-scatter-ABOUT-go-2SG
'you (SG) scatter betel refuse on the path as you go!'

(k) *makandəksenama*
    ma-kandək-se-ona-ma
    ITR.IMP-sleep-NIGHT-2SG-NEC
    'you (SG) must sleep!'

(l) *mumoran   kutuməndəmana*
    mə-moran   ku-t-mənd-ma-ona
    eat-thing TR.IMP-CAUS-finish-DUR-2SG
    'you (SG) continue to finish the food!'

(m) *ŋgakaketaməndə*   *mandakamben*
    ŋga-ka-ketam-nda   ma-nda-ka-mbe-n
    INV-FIRST-put.ashore-IMP.1.OBJ   ITR.IMP-NOW-FIRST-bear-1SG
    'put me ashore first and let me give birth now!'

(n) *moka*   *kena*   *ŋga*   *ŋgasambotəndə*
    ma-o-ka   ke-na   ŋga   ŋga-sambo-t-nda
    ITR.IMP-2SG 1DL-POSS DAT   INV-leave-APPL-IMP.1.OBJ
    'you (SG) go and leave us (DL)!'

(o) *nana* *ma*   *gadaerakotənd*   *ya*
    nana   ma   ŋga-nda-e-ra-kot-nd   ya
    mama   1SG   INV-NOW-come-stay-carry-IMP.1.OBJ   EMPH
    'mama, come and get me now!'

Examples (5.61m, n, o) are examples of inverse imperative verbs, all by definition with a local first person object. Inverse imperative verbs occur not only with the inverse prefix *ŋga*, but also with a suffix -*nd(a)* that indicates the local first person object but does not specify its number (compare (5.61m) with (5.61n). Note that in (5.61n), the verb is inverse with a first person object, yet that object is treated as a oblique constituent marked by the the dative postposition *ŋga*. Inverse inflection for imperatives, however, does not seem to be obligatory with local first person objects. Compare (5.61c, f) with (5.61m, n), though the former examples were elicited, while the latter are drawn from the narrative texts.

5.3.4.3 Prohibitives

Prohibitives are negative imperatives, that is, if an imperative means 'do X', a prohibitive means 'don't do X'. In Kopar prohibitives are expressed in two ways, one, analytic, with the negative desiderative verb *nda-* 'don't want' taking the prohibited action as a complement, and the other synthetic, via a bound verbal suffix -*arək* PROHIB. *nda-* 'don't want' like the negative *kay-* attracts bound pronominals from the verb of the complement, which again are drawn from the set for unrealized tenses and moods of Table 3:

(5.62) (a) mi arəm kiri ndana
mi arəm ki-ri nda-ona
2SG water bathe-DOWN PROHIB-2SG
'don't bathe!'

(b) awr təkari daduku
awr t-kari nda-onduku
fire CAUS-put PROHIB-2PC
'don't you (PC) start a fire!'

The synthetic prohibitives are formed with the suffix -*arək*, to which are appended the nominative pronominal agreement suffixes from Table 4, surprisingly a set normally limited to realized events, which prohibited events decidedly are not. Intransitive prohibitives lack the intransitive prefix *ma-*, but transitive ones can take the transitive prefix *ku-* (5.63e):

(5.63) (a) murarəkənaya
mur-arək-naya
afraid-PROHIB-2SG
'don't you (SG) be afraid!'

(b) nuŋgo si abeb warəkiya
nuŋgo si abeb o-arək-iya
very do only go-PROHIB-PC
'don't you (PC) go completely!'

(c) kandawarəkəmbaya
ka-nda-o-arək-mbaya
PERM-NOW-go-PROHIB-DL
'you (DL) can't go just now!'

(d) samerarəkəmbaya
samer-arək-mbaya
wait-PROHIB-DL
'don't you (DL) wait!'

(e) *nambrin   kusamaŋgarəkənaya*
nambrin   ku-samaŋ-arək-naya
eye   TR.IMP-choose-PROHIB-2SG
'don't you (SG) stare!'

(e) *kitarəkəna   mana   ŋga*
kit-arək-naya   ma-na   ŋga
hide-PROHIB-2SG   1SG-POSS   DAT
'don't hide them from me!'

(f) *nunon   karəkiya*
nunon   ka-arək-iya
thought   PERM-PROHIB-PC
'you (PC) don't have to think (about it)!'

Example (5.63f) is interesting in that there is no verb root involved (*nunon* 'thought, mind' is clearly a noun as witnessed by its final /n/) and the prohibitive suffix *-arək* is simply appended to the permissive prefix *ka-*. The combination of both a permissive and prohibitive affix in (5.63 c, f) is itself semantically unusual; the meaning is something like 'it's not possible now, but will be later'.

5.3.4.4 Hortatives

Hortatives are exhortations for the speaker or a group containing the speaker to do or be able to do an action; they translate as 'let' in English, as in 'let's go'. Hortatives are formed in the same way as imperatives, except that the bound pronominals are drawn from the first person series of Table 3 for unrealized tenses and moods. However, for first person singular subjects of transitive verbs, an alternative to the usual transitive imperative prefix *ku-* is to use the first person singular ergative prefix *na-* (5.64f). Some examples:

(5.64) (a) *paŋgə   mandaok*
paŋgə   ma-nda-o-okə
1PC   ITR.IMP-NOW-go-1PC
'let us (PC) go now'

(b) *maprembiya*
ma-pre-mbi-oya
ITR.IMP-die-IM.FUT-1SG
'let me die'

(c) ndata   makandəksemekə
    ndata   ma-kandək-se-ome-okə
    where   ITR.IMP-sleep-NIGHT-wh-1PC
    'where should we (PC) sleep?'

(d) manak   yowa ya   dabagarike
    ma-na-k   yowa ya   nda-mbaŋgari-oke
    1SG-POSS-NE   DIST EMPH NOW-kill-1DL
    'let us (DL) kill that (one) of mine now'

(e) ke   mandaŋgareke
    ke   ma-nda-ŋgara-e-oke
    1DL   ITR.IMP-NOW-front-come-1DL
    'let us (DL) now come in the front'

(f) karəkarək   ena-mb   karan
    karəkarək   ena-mb   karan
    pillow   PROX.SG-OBL   head
    nandateranəto
    na-nda-t-e-ra-n-to
    1SG.ERG-NOW-CAUS-come-stay-IMP-DEP
    'let me put my head on this pillow and'

Note in (5.64f) the distinct imperative suffix *-n ~ -na* that is again used here with first person. It is also exemplified in (5.56a, b) and (5.58a) with second and third person, though what determines the its usage and allomorphy is unclear.

5.4 Verb stem derivations: Valence changes

As we have seen, the concept of transitivity is very important in Kopar in determining the shape of verbal morphology. Both the choice of imperative prefix and the options for bound pronominals depend on the contrast between transitive and intransitive verbs. With some exceptions involving those that have cognate objects (see section 6.1.2.2), verb roots in Kopar are rather rigidly divided into transitive and intransitive classes. But the expression of events sometimes requires more and sometimes fewer participants than the transitivity of a verb root permits, and in such cases morphological derivations apply to add an extra core argument to an intransitive verb, producing now a transitive verb stem, or to delete a core argument from a transitive verb, resulting in an intransitive verb stem. Due to the short period of field work and restricted corpus, unfortunately the data on these processes are somewhat limited.

5.4.1 From transitive verb root to intransitive verb stem

5.4.1.1 Reflexivization
A reflexive construction is one in which a subject performs an action on himself or herself; hence the referents of both the subject and object are identical. Languages deal with this is certain ways; for example, in English by binding the coreferential object with a specialized pronouns series *PRO-self*. Other languages, in recognizing the equivalence of the referents of the subject and object, regard the verb as only having one core argument, i. e, it is an intransitive verb, so mark it as such as a derived intransitive. A transitive verb root that normally has two distinct core arguments, a subject and an object, now has only one, functioning simultaneously as both subject and object. Kopar has analogs of both of these options. When functioning in a reflexive construction, transitive verb roots occur with a prefix *ti-* REFL (*tu-* following *u-* 3SG or by vowel harmony with a following /u/) to mark reflexivization. Such verbs are formally intransitive verb stems and inflect as intransitive verbs. Consider the following contrastive examples:

(5.65) (a) *mu   mu   mbusamaytaŋgoya*
          mu   mu   mbu-samayt-aŋg-oya
          3SG 3SG 3.ERG-see-PRES-3SG
          'he/she$_1$ sees him/her$_2$.'

    (b) *mu   (mu-na-ndi)   utusamaytaŋgoya*
          mu   (mu-na-ndi)   u-ti-samayt-aŋg-oya
          3SG 3SG-POSS-ADV 3SG-REFL-see-PRES-3SG
          'he/she sees himself/herself'

Example (5.65a) is a normal transitive clause. The verb has transitive inflection with *mbu-* 3.ERG, and the two pronouns *mu* 3SG must be disjoint in reference. (5.65b) is the reflexive correspondent. The verb now has the prefix *ti-* REFL and has intransitive verb inflection with the prefix *u-* 3SG. The object pronoun may or may not be elided, but if realized, must now in a reflexive form with the suffix *-ndi*, normally an adverbial suffix meaning 'by', which requires the pronoun to take the possessive suffix *-na*. Here are examples in the other persons:

(5.66) (a) ma   mu   nasamaytaŋgoya
ma   mu   na-samayt-aŋg-oya
1SG   3SG   1SG.ERG-see-PRES-1SG
'I see him/her'

(b) ma   (ma-na-ndi)   matisamaytaŋgaya
ma   (ma-na-ndi)   ma-ti-samayt-aŋg-aya
1SS   1SG-POSS-ADV   1SG-REFL-see-PRES-1SG
'I see myself'

(5.67) (a) mi   mu   kusamaytəka
mi   mu   ku-samayt-ka
2SG   3SG   TR.IMP-see-SG.IMP
'you (SG) see him/her/it!'

(b) mi   (mi-na-ndi)   matisamaytəka
mi   (mi-na-ndi)   ma-ti-samayt-ka
2SG   2SG-POSS-ADV   ITR.IMP-REFL-see-SG.IMP
'you (SG) see yourself!'

In (5.66) and (5.67), the (a) examples are normal transitive clauses with subject and object core arguments, and the verbs have normal transitive inflection; note especially the transitive imperative prefix in (5.67a). The (b) examples are reflexives with the prefix *ti-*, and the verbs now inflect intransitively; again, note now the intransitive imperative prefix in (5.67b). Here is an example of reflexivization with verb incorporation, i. e. a serial verb construction:

(5.68) mambiturikatambiko
ma-mbi-ti-uri-kata-mbi-oko
ITR.IMP-AGAIN-REFL-say.name-speak-IM.FUT-2DL
'you (DL) call the names of yourselves (DL) again'

Note the whole verb is again intransitive, marked by the intransitive imperative marker *ma-*.

Contrasting with the reflexive constructions of (5.65)-(5.68), there is a second reflexive construction in Kopar that does not involve *ti-* REFL and detransitivization of the verb. This construction is a calque of Tok Pisin constructions involving the morpheme *yet* 'self', as in *mi yet wokim* 'I myself made it' or *mi lukim mi yet* 'I saw myself'. The Kopar word corresponding to *yet* is *raway* 'self', used in the non-reflexive exclusory meaning in (5.69):

(5.69) (a) mu   Inambaren raway okoya
           mu   Inambaren raway o-k-oya
           3SG   PN   self   go-FR.PAST-3SG
           'he, Inambaren himself, went'

     (b) mayn   raway   sirem   su,   nana   nda   yayan   aynde
         mayn   raway   sirem   su,   nana   nda   yayan   aynde
         husband   self   platform on   mama   and   papa   here
         'the husband himself on the platform, mama and papa here'

Reflexives with *raway* were obtained in elicitation sessions when the argument coreferential to the subject was a recipient, rather than a direct object. In *raway* 'self' reflexivization constructions the verbs are not detransitivized, but surprisingly the reflexivized recipient does not occur with the dative postposition *ŋga*; perhaps -*ndi* 'by' blocks that:

(5.70) (a) manandi   raway   ruaŋ   nasaytundukoya
           ma-na-ndi   raway   ruaŋ   na-sayt-ndək-oya
           1SG-POSS-ADV self   coconut   1SG.ERG-pick-NR.PAST-1SG
           'I picked a coconut for myself'

     (b) mi-na-ndi   raway   ruaŋ   isaytundukona
         mi-na-ndi   raway   ruaŋ   i-sayt-ndək-ona
         2SG-POSS-ADV   self   coconut   2-pick-NR.PAST-2SG
         'you (SG) picked a coconut for yourself'

5.4.1.2 The detransitivizer
The detransitivizer suffix converts a transitive verb root into an intransitive verb stem by suppressing the object grammatical function. It could be considered a type of antipassive morpheme. The object is completely eliminated, not simply demoted from being a core argument into being an oblique, as the antipasssive in some languages does. The form of the detransitivizer is -*am*, where the vowel is deleted following another non-high vowel. Consider the following examples:

(5.71) (a) rikam   kombar   yowa   kukatirəmbiko
           rikam   kombar   yowa   ku-katir-mbi-oko
           bamboo   two   DIST   TR.IMP-sever-IM.FUT-2DL
           'you (DL) cut those two pieces of bamboo!'

(b) *kindamorəmb ukatiramandə*
kindamorəmb u-katir-am-andə
bindings   3SG-sever-DETR-SEQ
'the bindings snapped and then'

(5.72) (a) *mbuniŋabuk*   *kay sur o*
mbu-ni-ŋga-mbu-k   kay sur o
3.ERG-put.inside-go.into-in.canoe-FR.PAST   canoe into
'he put it in the canoe'

(b) *maniŋombiko*
ma-ni-ŋga-o-am-mbi-oko
ITR.IMP-put.inside-go.into-go-DETR-IM.FUT-2DL
'you (DL) go inside!'

(c) *akən usianəŋrak*
akən u-si-anəŋra-k
sun   3SG-do-3PC.DAT-FR.PAST
ŋginiŋomәkiya
ŋgi-ni-ŋga-o-am-k-iya
3PL-put.inside-go.into-go-DETR-FR.PAST-PC
'the sun set on them (PC) and they (PC) went inside'

Example (5.72a) illustrates the verb in its transitive usage, while (5.72b, c) are derived intransitives with *-am* DETR. Note the overt intransitive imperative prefix in the detransitivized (5.72b), but the transitive imperative prefix in the fully transitive (5.71a).

A number of verbs denoting changes of state occur lexicalized with this suffix. These verbs are therefore inherently intransitive, so that when they function transitively they require derivation with the causative prefix (see section 5.4.2.1):

(5.73) (a) *mu   nambrin bubitupwaramandə*
mu   nambrin mbu-mbi-t-pwar-am-andə
3SG eye   3.ERG-AGAIN-CAUS-open-DETR-SEQ
'she opened her eye again and then'

(b) *mbuturәŋgamәkoya*   *kaynda*
mbu-t-rəŋg-am-k-oya   kay-onda
3.ERG-CAUS-wake.up-DETR-FR.PAST-3SG   NEG-3SG
'he didn't wake it up'

The change of position verb *kusa-* 'arrive, go out' is particularly interesting in that it can occur with (5.74b) or without (5.74a) the detransitivizing suffix, but when present, transitivization appears to require the causative prefix (5.74c):

(5.74) (a) *ukusanumbwakoya*
u-kusa-anumbwa-k-oya
3SG-arrive-3PL.DAT-FR.PAST-3SG
'it (a canoe) arrived at where they (PL) were'

(b) *ukusamenəmbakoya*
u-kusa-am-e-anəmba-k-oya
3SG-go.out-DETR-come-3DL.DAT-FR.PAST-3SG
'he went out and came up to them (DL)'

(c) *mbutukusaməkəto*   *indaimbot*
mbu-t-kusa-am-k-to   inda-imbot
3.ERG-CAUS-go.out-DETR-FR.PAST-DEP   house-nose
mbuturəməkəto
mbu-t-rəmə-k-to
3.ERG-CAUS-stand-FR.PAST-DEP
'he took her out and stood her up in the gable and'

5.4.2 From intransitive verb root to transitive verb stem

Crosslinguistically, the verbal derivations that increase the number of arguments of a verb, those that derive a transitive verb stem from an intransitive verb root, are causatives and applicatives (Comrie 1985). In the limited data available, a causative and two applicative derivational affixes have been determined.

5.4.2.1 Causatives

Causative constructions add a subject argument to the core argument array of a verb. In many languages such as Turkish (Comrie 1974) or Yimas (Foley 1991), causative derivations add another subject argument to a monotransitive verb root to produce a ditransitive verb stem. But it does not seem possible in Kopar to derive syntactically ditransitive verbs by the causative derivation. In Kopar, the causative prefix *t-* is added to verb roots to derive transitive verb stems, most commonly verb roots denoting states or changes of state (5.75a, b, c d, i), but not restricted to them (5.75e, f, g, h):

(5.75) (a) *mora   kandan   nor   mbutuprenaŋgoya*
mora   kandan   nor   mbu-t-pre-n-aŋ-oya
thing   sick   man   3.ERG-CAUS-die-LK-PRES-3SG
'sickness is killing the man'

(b) *akən   naŋgun   mana   pet   tumbutukama*
akən   naŋgun   ma-na   pet   t-mbu-t-kam-a
sun   skin   1SG-POSS   dark   PFV-3.ERG-CAUS-become-PFV
'the sun darkened my skin'

(c) *awr   arəm   ududuk   tumbutusya*
awr   arəm   ududuk   t-mbu-t-si-a
fire   water   hot   PFV-3.ERG-CAUS-become-PFV
'the fire heated the water'

(d) *moran   mbutundatukondu*
moran   mbu-t-ndat-k-ondu
thing   3.ERG-CAUS-know-FR.PAST-3PL
'they (PL) came to know the issue'

This seems a specialized use of the causative prefix with *ndat-* 'know', not 'cause someone to know' as expected, but 'come to know', 'become aware of' and thereby shares some of the meaning of the homophonous comitative applicative of section 5.4.2.2.1 below.

(e) *kutururujabiduku   paret   ŋgari*
ku-t-ruruɲja-mbi-onduku   paret   ŋgari
TR.IMP-CAUS-shake-IM.FUT-2PC   outside   DAT
'you (PC) shake (it) until (it comes) out'

(f) *ma   tamənd   ŋga   masak   sur   nambrin*
ma   tamənd   ŋga   masak   sur   nambrin
1SG   fish   D   ocean   inside   eye
natərəmaraŋgoya
na-t-rəmə-ar-aŋ-oya
1SG.ERG-CAUS-stand-PROG-PRES-1SG
'I'm fixing my eye on the fish in the ocean'

(g) *mbuturəmukoya*
mbu-t-rəmə-k-oya
3.ERG-CAUS-stand-FR.PAST-3SG
'he stood her up'

(h) *indan   mbuturikeranak*
    indan   mbu-t-riker-ana-k
    house   3.ERG-CAUS-get.up-3SG.DAT-FR.PAST
    'she erected a house for her'

(i) *naməŋ   mbutuparetukoya*
    naməŋ   mbu-t-paret-k-oya
    leg   3.ERG-CAUS-outside-FR.PAST-3SG
    'he pulled (his) leg out'

(j) *utunaɲjakoya*
    u-t-naɲja-k-oya
    3SG-CAUS-be.at.bank-FR.PAST-3SG
    'he kept to the river bank'

The prefix *t-* most probably derives from an older verb root *tu-* 'hit' in a serial verb construction that has undergone phonological reduction and re-analysis (Yimas has a cognate verb root *tu-* 'kill'). The verb root *kam-* 'become' in (5.75b) is almost certainly a borrowing from Tok Pisin *kamap* 'arrive, become'; the native root for the meaning is *si-* 'become, make, do, feel, happen', found in (5.75c), a frequent polysemy in Lower Sepik-Ramu languages.

5.4.2.2 Applicatives

An applicative derivation is the complement of a causative derivation. Causatives add a subject to the core arguments of a verb, while applicatives add an object core argument. Two probable applicatives have been identified in the Kopar corpus, a comitative and a general applicative, whose function is still somewhat obscure given the limited data available.

5.4.2.2.1 The comitative applicative

This applicative morpheme has the same form as the causative and almost certainly has arisen historically from that. To see how this could come about, consider the following example:

(5.76) *kiŋgep   mbuturaporaposarorokondu*
        kiŋgep   mbu-t-rapo+rapo-sa-ar-oro-k-ondu
        ladder   3.ERG-COM/CAUS-run.RED-IN-PROG-EXT-FR.PAST-3PL
        'they (PL) kept running in with a ladder'

Obviously, a ladder, being inanimate, cannot run around under its own steam. It needs to be caused to run. At the same time, it cannot be run around without someone or some people holding it while they run around with it; hence the ladder accompanies them while they are running with it. It appears that it was this use of the causative prefix with such motion verbs that is the source of the comitative meaning of this prefix. However, in many cases, it can no longer be analyzed as a plausible causative morpheme, but has now been re-analyzed transparently as a comitative applicative prefix, although the meaning is broader in some cases (5.77e, f, g):

(5.77) (a) ŋgatəraraŋgənaya
ŋga-t-ra-ar-aŋg-naya
INV-COM-stay-PROG-PRES-1SG
'she looks after me' (literally 'stays with me')

(b) mayndəpak mbuturarorokududu
mayndəpak mbu-t-ra-ar-oro-k-undundu
husband.PL 3.ERG-COM-stay-PROG-EXT-FR.PAST-3PC
'they (PC) remained with the husbands for a while'

(c) Wak yo mbutukararorokondu
Wak yo mbu-t-kar-ar-oro-k-ondu
PN DEF 3.ERG-COM-walk-PROG-EXT-FR.PAST-3PL
'they (PL) walked around with Wak for a while'

(d) nəmandəpak natəprəkaŋoya
nəmandəpak na-t-prək-aŋg-oya
woman.PL 1SG.ERG-COM-work-PRES-1SG
tənasambota
t-na-sambo-t-a
PFV-1SG.ERG-leave-APPL-PFV
'I work with the women, but I've left them behind'

(e) bubitərariekududu
mbu-mbi-t-rari-e-k-undundu
3.ERG-AGAIN-COM-cry-come-FR.PAST-3PC
'they (PC) cried over her again as they (PC) came'

(f) uprekəto mbuturarisekududu
u-pre-k-to mbu-t-rari-se-k-undundu
3SG-die-FR.PAST-DEP 3.ERG-COM-cry-NIGHT-FR.PAST-3PC
'he died, and they (PC) cried all night over him'

(g) kakandək   yo   mbutumuranak
    kakandək   yo   mbu-t-mur-ana-k
    older.same.sex.sibling DEF 3.ERG-COM-afraid-3SG.DAT-FR.PAST
    'the older sister was afraid of him'

The last three examples are problematic as to whether the prefix is properly analyzed as the comitative applicative or the causative (see also (5.75d). Causative morphemes introduce subjects, and while the dead person may be the causer of the wailing in (5.77e, f), he is not the subject, the wailers are. Applicatives introduce objects, and here the dead person is the reason or cause why the the wailers are crying; they are wailing because of him. Hence the analysis of this as the applicative seems better motivated than the causative, although the meaning clearly departs from the core meaning of comitative. This example again illustrates the likely origin of the comitative applicative prefix in the causative prefix. It is interesting to note that essentially the same meaning can be expressed with just a dative pronominal agreement suffix, arguing again that is more likely an applicative than a causative:

(5.78) mbirarianak
       mbi-rari-ana-k
       3DL-cry-3SG.DAT-FR.PAST
       'they (DL) cried over her'

Finally, in example (5.77g), the applicative seems to introduce a stimulus, a cause of the fear, again perhaps betraying the origin of this use of *t-* in the causative, but note again the stimulus is realized as a dative suffix.

5.4.2.2.2 The general applicative
The general applicative is still poorly understood, and the label applicative may not even be the best label for it. It seems to express that a transitive verb occurs with a highly contextually salient object or dative participant. Its form is *-t*, a suffix immediately following the verb root. Consider these examples with *sambo-* 'leave, put, abandon, deposit':

(5.78) (a) mbusambok
           mbu-sambo-k
           3.ERG-leave-FR.PAST
           'they left'

(b) *tiŋgi   mbekəmbaya   kayn   mbusambokondu*
    tiŋgi   mbi-e-k-mbaya   kayn   mbu-sambo-k-ondu
    behind   3DL-come-FR.PAST-3DL   canoe   3.ERG-leave-FR.PAST-3PL
    'they (DL) came later; they (PL) had left a canoe'

(c) *mbusamborerutakondu*
    mbu-sambo-rer-o-ta-k-ondu
    3.ERG-leave-scatter-go-OUT-FR.PAST-3PL
    'they (PL) left her and turned around'

(d) *mbusambotək*
    mbu-sambo-t-k
    3.ERG-leave-APPL-FR.PAST
    'they abandoned her'

(e) *nəmandəpak   natəprəkaŋgoya*
    nəmandəpak   na-t-prək-aŋg-oya
    woman.PL   1SG.ERG-COM-work-PRES-1SG
    tənasambota
    t-na-sambo-t-a
    PFV-1SG.ERG-leave-APPL-PFV
    'I work with the women, but I've left them (PL) behind'

(f) *punduŋ   sur   o   mbusambotanak*
    punduŋ   sur   o   mbu-sambo-t-ana-k
    lap   on   3.ERG-put-APPL-3SG.DAT-FR.PAST
    'she put him on her lap'

(g) *ara   ŋga   apən   isambotəmananamena*
    ara   ŋga   apən   i-sambo-t-ma-anana-ome-ona
    what   DAT   food.bits   2-leave-APPL-DUR-2SG.DAT-*wh*-2SG
    'why do you (SG) keep leaving food bits of yours'

In the first three examples, *sambo-* 'leave, put, abandon, deposit' occurs without the general applicative suffix *-t*, and the objects of the verb are either left unspecified (5.78a), inanimate and indefinite (5.78b) or of lower salience in context (5.78c). In examples (5.78d-g), the verb root is derived with the general applicative suffix. In each case the object is of much higher salience, hence the translation of 'abandon' in (5.78d) as opposed to just 'leave' in (5.78a, c). In (5.78e, f), the objects of the verb are human and of high salience in context, and while the object in (5.78g) is inanimate, in the context of the story it is a highly salient prop, which risks revealing the protoganist's intentions.

Another verb that behaves like *sambo-* 'leave, put, abandon, deposit' is *tadaba-* 'prepare, fix, make right' (Tok Pisin *stretim*):

(5.79) (a) *moran   mbutadabanəŋgrak*
moran   mbu-tandamba-anəŋgra-k
thing   3.ERG-fix-3PC.DAT-FR.PAST
'they prepared food for them (PC)'

(b) *sara-n   mbutadabatukududu*
sara-n   mbu-tandamba-t-k-undundu
report-NMLZ   3.ERG-fix-APPL-FR.PAST-3PC
'they (PC) straightened out the report'

In both examples the verb is inflected transitively with *mbu-* 3.ERG. But in (5.79a) the object is backrounded, of generic/indefinite information, and the verb appears without *-t* APPL. In (5.79b), getting the story straight in order to tell it to the women is the whole point of the text at this point; *saran* 'story, report' is a highly salient object and so the verb takes *-t* APPL.

The verb root *kadəb-* 'wash, clean' is one like *mə-* 'eat' that seems to be ambitransitive; it is inflected intransitively in (5.80a) and transitively in (5.80b), but with the presence of *-t* APPL in the latter:

(5.80) (a) *ma   mora   makadəbəndəkənaya*
ma   mora   ma-kandəmb-ndək-naya
1SG   thing   1SG-wash-NR.PAST-1SG
'I washed thing(s)'

(b) *ma   mora   yo   nakadəbətundukoya*
ma   mora   yo   na-kandəmb-t-ndək-oya
1SG   thing   DEF   1SG.ERG-wash-APPL-NR.PAST-1SG
'I washed the thing'

However, like *sambo-* 'leave, put, abandon, deposit' and *tadaba-* 'prepare, fix, make right', *kadəb-* 'wash, clean' can be used transitively without *-t* APPL as well; compare (5.81a) with (5.81b):

(5.81) (a) *napar   mbukadəbundukoya*
napar   mbu-kandəmb-ndək-oya
hand   3.ERG-wash-NR.PAST-3SG
'he/she washed (his/her) hand'

(b) *napar mbukadəbətundukoya*
   napar mbu-kandəmb-t-ndək-oya
   hand 3.ERG-wash-APPL-NR.PAST-3SG
   'he/she washed (his/her) hand'

There is one case in the narrated texts, though, where -*t* APPL does seem to function more like a canonical applicative, but again has anomalous features:

(5.82) (a) *maprembiya*
   ma-pre-mbi-oya
   ITR.IMP-die-IM.FUT-1SG
   'let me die'

(b) *awr upretanəmbak*
   awr u-pre-t-anəmba-k
   fire 3SG-die-APPL-3DL.DAT-FR.PAST
   'the fire died on them (DL)'

The verb root *pre-* 'die' is intransitive and inflected as such in (5.82a). In (5.83b) the general applicative suffix is employed and an argument is added, but it appears as a dative pronominal agreement suffix, and the verb is still inflected intransitively with *u-* 3SG instead of the expected *mbu-* 3.ERG (however, the possibility that the suffix is actually -*ta* OUT cannot be entirely excluded, i. e. 'the fire died out/ away on them (DL)'). Of course, an intransitive inflectional pattern is typical of uncontrolled events with dative affixes, as discussed in (5.2.4). They do not function as grammatical objects, but that does raise even more questions about the true function of -*t* APPL. Unfortunately, given the data available, no further clarification can be provided, but it should be pointed out that many of the most canonical, highly transitive verbs end in /t/, indicating perhaps lexicalization of this suffix: *nambrat-* 'spear', *kumbrat-* 'break', *kot-* 'carry', while 'see' is lexicalized with two variants, *subo-* and *samayt-*, the latter with a final /t/.

5.5 Verb theme derivations

Verb themes are derived from verb stems in the same way that verb stems are derived from verb roots: by the specification of additional morphological verbal categories. While the Kopar verb may not be as obviously hierarchical as that of Yimas (Foley 1991:353–369), the distinction between verb stem and verb theme is still descriptively useful for it. It is these additional verbal categories that are responsible for the classification of Kopar as a moderately polysynthetic language, in particular, incorporation. Theme derivational processes differ from stem derivational processes in that they seem more 'inflectional', modulating the basic meaning of the verb by adding specifications for possessor, time, manner and direction of action, rather than re-structuring the grammatical functions of its core arguments.

5.5.1 Possessor raising

Possessor raising was never encountered in elicitation sessions, but it is a very common feature of the narrative texts. Possessor raising occurs when a human possessor of a noun occurs as a dative pronominal agreement suffix instead of or in addition to (5.83f) being realized as a -na POSS marked constituent in a noun phrase. Nouns from which possessors can be raised are most commonly body parts, although they are not limited to those in Kopar, but can be kin terms or other types of salient possessed nouns (5.83c, 5.84c, e). Furthermore, unlike many other languages in which possessor raising is limited to core arguments, subjects or objects, possessors can be raised from postpositional phrases in Kopar.

Here are some examples of possessor raising from subjects. Possessor raising from subjects appears limited to subjects of formally intransitive, mostly unaccusative, verbs:

(5.83) (a) *nambrin ukusamanak*
nambrin u-kusa-am-ana-k
eye 3SG-go.out-DETR-3SG.DAT-FR.PAST
'his eye goes out' = 'he looks out'

(b) *nambrin usirianak*
nambrin u-siri-ana-k
eye 3SG-go.down-3SG.DAT-FR.PAST
'his eye went down' = 'he looked down'

(c) *nanan urikeranakoya   miɲjir*
nanan u-riker-ana-k-oya   miɲjir
mama 3SG-get.up-3SG.DAT-FR.PAST-3SG urine
uprasenakoya
u-pra-se-ana-k-oya
3SG-excrete-NIGHT-3SG.DAT-FR.PAST-3SG
'her mother got up and she urinated during the night'

134 — Chapter 5 Verbal morphology

    (d) *indaimbot nambrin urukoranakoya*
        inda-imbot nambrin u-rukor-ana-k-oya
        house-nose eye 3SG-go.ashore-3SG.DAT-FR.PAST-3SG
        'her eye went ashore (toward) the gable' = 'she looked toward the gable'

    (e) *naŋgun munak yo*
        naŋgun mu-na-k yo
        skin 3SG-POSS-NE DEF
        ukusaməmanakoya
        u-kusa-am-ma-ana-k-oya
        3SG-go.out-DETR-DUR-3SG.DAT-FR.PAST-3SG
        'his skin kept coming off'

Possessor raising is also common from the object noun phrase of transitive verbs:

(5.84) (a) *namən mburatanəmbakoya*
        namən mbu-rat-anəmba-k-oya
        leg 3.ERG-stamp-3DL.DAT-FR.PAST-3
        'they stamped their (DL) feet'

    (b) *namən mbwerakotanak*
        namən mbu-e-ra-kot-ana-k
        leg 3.ERG-come-stay-carry-3SG.DAT-FR.PAST
        'she brought her leg'

    (c) *ara ŋga apən*
        ara ŋga apən
        what DAT food.bits
        isambotmananamena
        i-sambo-t-ma-anana-ome-ona
        2-leave-APPL-DUR-2SG.DAT-*wh*-2SG
        'why do you (SG) keep leaving food bits of yours (SG)?'

    (d) *puruŋ ŋga napar mbutukaməmanak*
        puruŋ ŋga napar mbu-t-kam-ma-ana-k
        betelnut DAT hand 3.ERG-CAUS-arrive-DUR-3SG.DAT-FR.PAST
        'he kept putting out his hand for betelnut'

    (e) *uren mbubagarianak*
        uren mbu-mbaŋgari-ana-k
        dog 3.ERG-kill-3SG.DAT-FR.PAST
        'they killed her dog'

Possessor raising from subjects of intransitive, particularly unaccusative, verbs and objects of transitive verbs is common among the languages of the world. What is unusual about Kopar is that it permits possessor raising from oblique constituents, even the complements of postpositional phrases:

(5.85) (a) *asumb sur mbusurorarianakoya*
asumb sur mbu-sur-o-ra-ri-ana-k-oya
mouth in 3.ERG-inside-go-stay-DOWN-3SG.DAT-FR.PAST-3SG
'he jammed it down into her mouth'

(b) *kambowen asumb o ndək ŋgiparianak*
kambowen asumb o ndək ŋgi-pari-ana-k
spines mouth PURP 3PL-extract-3SG.DAT-FR.PAST
'they removed the spines from his mouth'

(c) *nimbep napar sur o ndək urotanakəto*
nimbep napar sur o ndək u-rot-ana-k-to
spear hand inside PURP 3SG-fall-3SG.DAT-FR.PAST-DEP
'the spear fell from his hand and'

(d) *pundun sur o mbusambotanak*
pundun sur o mbu-sambo-t-ana-k
lap on 3.ERG-put-APPL-3SG.DAT-FR.PAST
'she put him on her lap'

5.5.2 Incorporation

Incorporation is regarded by Foley (2017b) to be the most important diagnostic for classifying a language as polysynthetic, and on this criterion Kopar does quality. The incorporation processes of Kopar seem to be rather less extensive than in some other polysynthetic languages like Onondaga (Woodbury 2018), Pawnee (Parks 1976; Cruikshank 2011), Nuu Chah Nulth (Davidson 2002), Nivkh (Mattissen 2003) or Yimas (Foley 1991), though this generalization could be a function of the limited period of fieldwork and the small corpus of data. Examples of incorporation are not easily elicited, and the running text material comprises no more than three narrative texts of moderate length and one shorter one, not all fully analyzable, not enough to provide copious examples of diverse types.

Incorporation is an older, more established term for what Mattissen (2003) calls dependent-head synthesis. It is a grammatical feature in which modifiers and satellites of a verb, like temporals, locationals/directionals, and manner adverbials, instead of appearing as independent words, as they would in a language like English, German or Swahili, are realized as bound affixes within the verb. In some more extreme polysynthetic languages like Onondaga (Woodbury 2018) even core arguments of a verb like transitive objects and sometimes intransitive subjects can be realized as bound morphemes within their governing verb. This is noun incorporation and this is not productive in Kopar; there are only a few highly lexicalized examples of this in the corpus. Productive incorporation in Kopar is restricted to adverbial notions like time, manner and direction, as well as exuberant verb incorporation. The morphemes expressing these notions are clearly formally incorporated because they occur within the sequence of morphemes initiated and terminated by bound pronominals.

5.5.2.1 Incorporation of temporals

As discussed in section 3.8, Kopar has a set of temporal words that are used to divide and label the different parts of the day:

tumbunan	'morning' (5am-9am)
sinan	'middle of the day' (9am-4pm)
wakənan	'late afternoon' (4pm-7pm)
rakamdan	'night' (7pm-5am)

Corresponding to each one of these time periods is a bound suffix of the same meaning that can be used instead of or in combination with the independent temporal word:

Table 11: Incorporated Temporal Suffixes.

-siri	'morning' (-si in the Wongan dialect)
-ar	'middle of the day'
-ma	'afternoon'
-se	'night'

These suffixes occur after the verb stem, i. e. after verb stem derivational suffixes like the detransitivizer and the general applicative, and before the aspectual suffixes. Note the formal similarity of *-ar* 'middle of the day' and *-ma* 'afternoon' to the aspectual suffixes *-ar* PROG and *-ma* DUR respectively and most likely the latter's ultimate source. The question, of course, arises is whether these are really just the same suffixes rather than homophonous ones. However, they are clearly synchronically distinct morphemes, as these aspectuals and temporals can co-occur:

(5.86) ambisen ukandəksemakoya minjir pra ŋga
 ambisen u-kandək-se-ma-k-oya minjir pra ŋga
 daughter 3SG-sleep-NIGHT-DUR-FR.PAST-3SG urine excrete DAT
 usianakoya
 u-si-ana-k-oya
 3SG-do-3SG.DAT-FR.PAST-3SG
 'the daughter was sleeping during the night and felt like urinating'

Here are a few examples of the use of these incorporated temporals:

(5.87) (a) ukararaesenumbwak
 u-karara-e-se-anumbwa-k
 3SG-ask-come-NIGHT-3PL.DAT-FR.PAST
 'during the night he asked them (PL) to come'

 (b) ŋgipunumak akən
 ŋgi-punu-ma-k akən
 3PC-work.sago-AFTERNOON-FR.PAST sun
 usianəŋrak 'paŋgə ndata
 u-si-anəŋra-k 'paŋgə ndata
 3SG-do-3PC.DAT-FR.PAST 1PC where
 makandəksemekə'
 ma-kandək-se-ome-okə'
 ITR.IMP-sleep-NIGHT-wh-1PC
 'they (PC) worked sago during the afternoon and the sun went down on them (PC): 'where should we (PC) sleep during the night?'

 (c) paŋgə inda kupa-mb ikandəksenaŋgiya
 paŋgə inda kupa-mb i-kandək-se-n-aŋg-iya
 1PC house big-OBL 1-sleep-NIGHT-LK-PRES-PC
 'we (PC) sleep in a big house'

 (d) ara mbusiaraŋgomendu
 ara mbu-si-ar-aŋg-ome-ondu
 what 3PL.ERG-do-DAY-PRES-wh-3PL
 'what are they (PL) doing?' (spoken during daytime)

 (e) mi ara-moran isisenaŋgomena
 mi ara-moran i-si-se-n-aŋg-ome-ona
 2SG what-thing 2-do-NIGHT-LK-PRES-wh-2SG
 'what are you (SG) doing?' (spoken at nighttime)

(f) *ŋgiramakəto,*   *nda wakəna nana*
ŋgi-ra-ma-k-to,   nda wakəna nana
3PC-stay-AFTERNOON-FR.PAST-DEP and afternoon mama
e   ŋga   usianakoya
e   ŋga   u-si-ana-k-oya
come DAT 3SG-do-3SG.DAT-FR.PAST-3SG
'they (PC) stayed until afternoon and afternoon mama wanted to come'

(g) *mu-moran   ŋgisik   ŋgimək*
mə-moran   ŋgi-si-k   ŋgi-mə-k
eat-thing   3PL-do-FR.PAST   3PL-eat-FR.PAST
ŋgiramak   ŋgirasek
ŋgi-ra-ma-k   ŋgi-ra-se-k
3PL-stay-AFTERNOON-FR.PAST   3PL-stay-NIGHT-FR.PAST
ŋgikandəkək
ŋgi-kandək-k
3PL-sleep-FR.PAST
'they (PL) made food, they (PL) ate, they (PL) stayed through the afternoon, they (PL) stayed until night, they slept'

(h) *nana yo   mora kanda sisirik*
nana yo   mora kanda si-siri-k
mama DEF thing sick   feel-MORNING-FR.PAST
aymbor payndəp kandəksirikoya
aymbor payndəp kandək-siri-k-oya
hearth beside sleep-MORNING-FR.PAST-3SG
'mama felt sick in the morning, and she slept by the side of the hearth'

5.5.2.2 Incorporation of adverbials

Obligatorily incorporated adverbials are a clear diagnostic feature of polysynthetic languages, especially in the languages of the Sepik region (Foley 2017b). So far, four incorporated adverbials have been identified in the small Kopar corpus; almost certainly there are more in the language. In Alamblak of the Sepik family (Bruce 1984), for example, there are a few dozen. Incorporated adverbials are prefixes that occur before the verb stem, i. e, before verb stem derivational prefixes like the causative/comitative or reflexive, and after any pronominal agreement prefixes.

5.5.2.2.1 nda- 'now'

The prefix *nda-* means 'right then, just now, at this/that moment'; it is very common in imperatives and hortatives. It is transparently related to *ndesa* 'today', so could be regarded as another example of an incorporated temporal, except that it is a prefix in the same position as other incorporated adverbials, not a suffix like the incorporated temporals. Some examples:

(5.88) (a) *mandasirinato*
ma-nda-siri-ona-to
ITR.IMP-NOW-go.down-2SG-DEP
'you (SG) go down now and'

(b) *saran   ndaerakotondukona*
sara-n   nda-e-ra-kot-onduk-ona
report-NMLZ   NOW-come-stay-carry-FUT-2SG
'you (SG) will get a report (gossip about you) now'

(c) *karəkarək   enamb   karan*
karəkarək   ena-mb   karan
pillow   PROX.SG-OBL   head
nandateranəto
na-nda-t-e-ra-n-to
1SG.ERG-NOW-CAUS-come-stay-IMP-DEP
'let me lay my head on this pillow now and'

(d) *manak   yowa ya   dabagarike*
ma-na-k   yowa ya   nda-mbaŋari-oke
1SG-POSS-NE   DIST EMPH   NOW-kill-1DL
'let us (DL) kill that (one) of mine now'

(e) *kandawarəkəmbaya*
ka-nda-o-arək-mbaya
PERM-NOW-go-PROHIB-2DL
'you (DL) can't go just now!'

(f) *mandaoke*
ma-nda-o-oke
ITR.IMP-NOW-go-1DL
'let us (DL) go now'

This particular incorporated adverbial occurs in an unusual construction only attested with verbs in the near past and present tenses and with a specialized meaning of 'almost, nearly'. Consider the following elicited examples:

(5.89) (a) ma mu nuŋgo mindi nadabagarindək
ma mu nuŋgo mindi na-nda-mbaŋgari-ndək
1SG 3SG very ? 1SG.ERG-NOW-kill-NR.PAST
'I almost killed him/her yesterday'

(b) ma mu nuŋgo mindi nadabagariaŋ
ma mu nuŋgo mindi na-nda-mbaŋgari-aŋ
1SG 3SG very ? 1SG.ERG-NOW-kill-PRES
'I'm almost killing him'

(c) mi mu nuŋgo mindi idabagarindukona
mi mu nuŋgo mindi i-nda-mbaŋgari-ndək-ona
2SG 3SG very ? 2-NOW-kill-NR.PAST-2SG
'you (SG) nearly killed him'

(d) mu muna ŋga nuŋgo mindi ruaŋ
mu mu-na ŋga nuŋgo mindi ruaŋ
3SG 3SG-POSS DAT very ? coconut
budaukindək
mbu-nda-uki-ndək
3.ERG-NOW-buy-NR.PAST
'he/she nearly bought a coconut for him/her'

The verb *bagari* 'kill' in (5.89a-c) is likely related to Tok Pisin *bagarap* 'be ruined, destroyed', though it should be analyzed underlyingly as *mbaŋgari*, given that it triggers denasalization in examples like (5.88d) and (5.89) and other examples in this monograph, though there are exceptions where it fails to do so (e.g. (5.37f) and (5.84e). There is, however, a verb root *baga-* 'kill' in Murik, and prenasalized voiced stops in Kopar correspond to plain voiced stops in Murik (see Appendix 1), so the Kopar verb could be a blend of an original verb *mbaŋga-* and Tok Pisin *bagarap*, but in any case an underlying form of *mbaŋgari-* is well motivated. These constructions lack the bound pronominal suffix *-oya* 1SG or 3SG in examples (5.89a, b, d). This results in the final consonant cluster /ŋg/ in (5.89b) surfacing after the present tense suffix *-aŋg* is orphaned without its usual following bound pronominal (again this is unusual because in such situations *-aŋg* is usually realized as *-a*). The incorporated temporal *nda-* means 'right then, at that moment' in these examples. But what has occurred at that moment is the failure of the action to complete. The exactness of the moment of failure, 'very nearly', is also emphasized by *nuŋgo* 'very', otherwise used with adjectival verbs (section 3.3). But the notion of the failure of the action, the 'almost' component, must come from the unglossed word *mindi*. This looks very suspiciously like a contraction of the verb root *mənd-* 'be finished' plus the adverbial suffix, *-ndi*, i. e *mənd-ndi* > *mindi*, but the semantics seems strange: how would something literally meaning 'very finished' end up meaning 'almost'?

5.5.2.2.2 *ŋga-* ~ *ka-* 'first'

This incorporated adverbial is transparently derived from the independent noun *ŋgara(n)* 'front, first place'. The *ka-* allomorph always occurs when preceding a syllable containing /k/, but is not restricted to nor required in that environment. Some examples:

(5.90) (a) *maŋgaramaka*
ma-ŋga-ra-ma-ka
ITR.IMP-FIRST-stay-DUR-2SG
'you (SG) continue to stay first!'

(b) *sapikindi   maŋgaebiduku*
sapiki-ndi   ma-ŋga-e-mbi-onduku
good-ADV   ITR.IMP-FIRST-come-IM.FUT-2PC
'it's good that you (PC) come first!

(c) *ko   maŋgaombiko   arəm*
ko   ma-ŋga-o-mbi-oko   arəm
2DL   ITR.IMP-FIRST-go-IM.FUT-2DL   water
maŋgakirimbiko
ma-ŋga-ki-ri-mbi-oko
ITR.IMP-FIRST-bathe-DOWN-IM.FUT-2DL
'you (DL) go first now and bathe!'

(d) *ŋgakaketamənda*
ŋga-ka-ketam-nda
INV-FIRST-put.ashore-IMP.1.OBJ
'you (SG) put me ashore first'

5.5.2.2.3 *pa-* 'still, yet

This incorporated prefix can overlap in the present tense and imperative with the meaning of *nda-* 'now', but has a wider range in meaning that an event or action is yet ongoing in any time period (see example 5.91b in the present tense). It is not all that common, but is attested a few times in the corpus (note the occurrence of two incorporated adverbials in (5.91a):

(5.91) (a) *mapaŋgarikerəka*
ma-pa-ŋga-riker-ka
ITR.IMP-STILL-FIRST-get.up-2SG
'you (SG) get up right now!'

(b) *ŋgu paomanaŋi*
ŋgu pa-o-ma-n-aŋg-iya
2PC STILL-go-AFTERNOON-LK-PRES-PC
'you (PC) are still going in the afternoon?'

(c) *ambisen upararisenakoya*
ambisen u-pa-rari-se-ana-k-oya
daughter 3SG-STILL-cry-NIGHT-3SG.DAT-FR.PAST-3SG
'the daughter was still crying for her during the night'

(d) *kubajərəm yowa uparotaranumbwak*
kubajərəm yowa u-pa-rot-ar-anumbwa-k
feathers DIST 3SG-STILL-fall-PROG-3PL.DAT-FR.PAST
'those feathers were still falling from them (PL)'

5.5.2.2.4 *mbi-* 'again'
The prefix *mbi-* 'again' means to repeat an action and is very common; it is very often subject to denasalization and triggers mutual denasalization of any preceding third person prefixes:

(5.92) (a) *gibek*
ŋgi-mbi-e-k
3PL-AGAIN-come-FR.PAST
'they came again'

(b) *mambikisabiduku*
ma-mbi-kisa-mbi-onduku
ITR.IMP-AGAIN-hide-IM.FUT-2PC
'you (PC) go hide again!'

(c) *imbiyaraŋənane*
i-mbi-e-ar-aŋg-na-n-e
2-AGAIN-come-PROG-PRES-2SG-LK-Q
'are you (SG) coming back?'

(d) *bubitərariekududu*
mbu-mbi-t-rari-e-k-undundu
3.ERG-AGAIN-COM-cry-come-FR.PAST-3PC
'they (PC) cried over her as they (PC) came back'

(e) *bubitokondu*
mbu-mbi-t-o-k-ondu
3.ERG-AGAIN-CAUS-go-FR.PAST-3PL
'they (PL) took him again'

5.5.2.3 Incorporation of directionals

Just as the incorporated temporals express the temporal coordinates of a reported event, incorporated directionals indicate the spatial coordinates of an event, either with respect to the place of the speech act or that of other events in the discourse. Incorporated directionals are a common feature of the more morphologically rich languages of the Sepik region; for a discussion of the situation in Kopar's sister language Yimas, see Foley (1991:346–353). The system in Kopar may not be as elaborated as Yimas, but that could simply be due to the restricted corpus. Directionals were never obtained in elicitation sessions, though they occur frequently enough in the narrative texts.

From the corpus, five directional suffixes as in Table 12 have been determined for Kopar; these appear after the verb stem with any detransitivizer or general applicative suffix, but before any incorporated aspectual or dative suffixes (5.93a, f, i, j):

Table 12: Incorporated Directional Suffixes.

-ri	'down, downriver'
-na	'up, upriver'
-sa	'in'
-ta	'out, away'
-am	'around, about'

(5.93) (a) *asumb sur mbusurorarianakoya*
asumb sur mbu-sur-o-ra-ri-ana-k-oya
mouth in 3.ERG-inside-go-stay-DOWN-3SG.DAT-FR.PAST-3SG
'he jammed it down into her mouth'

(b) *iniŋ ŋgitonak*
iniŋ ŋgi-t-o-na-k
stone 3PL-CAUS-go-UP-FR.PAST
'they (PL) lifted the stone up'

(c) *mbionakəmbaya*
mbi-o-na-k-mbaya
3DL-go-UP-FR.PAST-3DL
'they (DL) went upriver'

(d) *suŋgandur mbukarisaok*
suŋgandur mbu-kari-sa-o-k
grass 3.ERG-put-IN-go-FR.PAST
'he put it inside the grass'

(e) *moran mbuturukorəsakududu*
moran mbu-t-rukor-sa-k-undundu
thing 3.ERG-CAUS-go.ashore-IN-FR.PAST-3PC
'they (PC) brought everything in ashore'

(f) *mayndəpak yo kiŋgep*
mayndəpak yo kiŋgep
man.PL DEF ladder
mbuturaporaposarorokondu
mbu-t-rapo+rapo-sa-ar-oro-k-ondu
3.ERG-COM-run.RED-IN-PROG-EXT-FR.PAST-3PL
'the men kept running in with a ladder'

(g) *ŋgirapotak*
ŋgi-rapo-ta-k
3PL-run-OUT-FR.PAST
'they ran out'

(h) *nəmandəpak numotanda mbutetakududu*
nəmandəpak numotanda mbu-t-e-ta-k-undundu
woman.PL man.PL 3.ERG-CAUS-come-OUT-FR.PAST-3PC
'the women brought the men out'

(i) *mayn aratək ukusatanumbuk*
mayn aratək u-kusa-ta-anumbu-k
husband good 3SG-go.out-OUT-3PL.DAT-FR.PAST
'the man came out to them (PL) well' (was in full view)

(j) *nəmbren okumbi mbutetanak*
nəmbren okumbi mbu-t-e-ta-ana-k
pig PL 3.ERG-CAUS-come-OUT-3SG.DAT-FR.PAST
'she brought the pigs out for him'

(k) *porakay yo tak kureramoka*
    porakay yo tak ku-rer-am-o-ka
    path DEF betel.refuse TR.IMP-scatter-ABOUT-go-2SG
    'you (SG) scatter betel refuse about on the path as you go!'

(l) *inaŋ mbusuroraməkondə,*
    inaŋ mbu-sur-o-ra-am-k-ondə
    oar 3.ERG-inside-go-stay-ABOUT-FR.PAST-3DL
    'they (DL) tossed the oars overboard'

5.5.2.4 Incorporation of nouns

Like its sister language Yimas (Foley 1991:319–321, 2017b), noun incorporation is not productive in Kopar. It seems to be highly lexicalized, and even so, in the overall corpus, it is very rare. There are only two probable examples in the corpus of noun incorporation:

(5.94) (a) *ke mandaŋgareke*
           ke ma-nda-ŋgara-e-oke
           1DL ITR.IMP-NOW-front-come-1DL
           'let us (DL) now come in front'

      (b) *normot mbutukaranəmeritak*
          normot mbu-t-karan-meri-ta-k
          man.PL 3.ERG-CAUS-head-follow-OUT-FR.PAST
          'he sent the men out' (literally 'caused to follow the heads')

The incorporated noun *ŋgara(n)* means 'front, forehead, forward section of a canoe'. In (5.94a) it is used in this last meaning, so 'let us come forward, in the front of a canoe' (the term 'forward' is also used in English to describe the front section of a ship, like 10 Forward in *Star Trek: The Next Generation*). In (5.94b), 'head-follow' is an idiom in Kopar, meaning a group moving in a straight line, behind the one in front'. Here it is causativized so means 'he sent them out moving in a straight line'.

5.5.2.5 Incorporation of verbs: Verb serialization

In an earlier well known theoretical and typological study of incorporation (Baker 1988), verb incorporation was essentially discussed with regard to causative and applicative constructions. While this may apply to Kopar in some cases; for example, the causative prefix is arguably derived from an earlier verb root *tu-* 'hit', still attested as part of the synchronic verb *tumanəŋ-* 'hit' and witnessed by the Yimas cognate *tu-* 'kill', in the sense I mean it here it refers something much wider. Verb incorporation here refers to the juxtaposition of more than one verb root or stem in a single inflected verb. This process is widespread in Kopar and weakly constrained, often described as verb serialization as I did for Yimas (Foley 1991) or verb compounding. The semantic relationship between the verb roots or stems is not indicated, so formally verb incorporation typically consists of simple juxtaposition. The temporal relationships between them can be such that the events described by each root or stem are simultaneous or in sequence; in the latter case, their linear order is iconic to the sequence of events. These constructions are properly called verb incorporation because an otherwise structure of two or more clauses is telescoped to one, by incorporating the verb roots into one complex verb theme. This then takes theme level derivations like temporal or adverbial incorporation or possessor raising as a whole, as well as full verbal inflections like tense or pronominal agreement affixes. Consider the conjoined clauses in (5.95a) and the incorporated verb structure in (5.95b)

(5.95) (a) *utunaɲjak* nda *urukorək*
u-t-naɲja-k nda u-rukor-k
3SG-CAUS-be.at.bank-FR.PAST and 3SG-go.ashore-FR.PAST
'he kept to the bank and went ashore'

(b) *umbitənaɲjarukorukoya*
u-mbi-t-naɲja-rukor-k-oya
3SG-AGAIN-CAUS-be.at.bank-go.ashore-FR.PAST-3SG
'he kept to the bank and went ashore again'

In the conjoined clauses structure of (5.95a) each verb takes its own subject pronominal agreement prefix and tense suffix, but in (5.95b), the complex incorporated verb takes the full verbal inflections, incorporated adverbial, subject agreement and tense, only as a whole.

Incorporated verb constructions can have up to three verb roots incorporated (more than that has not been found in the corpus), and the temporal relationships between the verb roots can be simultaneous or sequential. The most common type of simultaneous incorporated verb constructions is those expressing associated motion (Guillaume and Koch 2021), in which an action is being performed while the performer is in motion to or from a location. These are most commonly formed with the basic motion verbs *e-* ~ *ya-* 'come' and *o-* ~ *wa-* 'go', but other verbs denoting movement such as *kusa-* 'go out' or *siri-* 'go down(river)' also occur. Some examples:

(5.96) (a) *tamənd mbiruesirik*
tamənd mbi-ru-e-siri-k
fish 3DL-shoot-come-go.down-FR.PAST
'they (DL) were spearing fish as they came downriver'

(b) *maraposawaraŋgaya*
ma-rapo-sa-o-ar-aŋg-aya
1SG-run-IN-go-PROG-PRES-1SG
'I am running inside (away from here)'

(c) *mbuturərəkwarəkududu*
mbu-t-rərək-o-ar-k-undundu
3.ERG-CAUS-push.(rək- RED)-go-PROG-FR.PAST-3PC
'they (PC) were shoving it as they (PC) went'

(d) *bubitərariekududu*
mbu-mbi-t-rari-e-k-undundu
3.ERG-AGAIN-COM-cry-come-FR.PAST-3PC
'they (PC) cried over her as they came back'

(e) *ŋgikumoyosekəŋgaya*
ŋgi-kumoy-o-se-k-ŋgaya
3PL-fish.with.net-go-NIGHT-FR.PAST-3PL
'they (PL) were going along fishing with a net at night'

But simultaneous incorporated verb constructions are not limited to associated motion:

(5.97) (a) *mbukaɲjimorakondu*
mbu-kaɲji-mora-k-ondu
3.ERG-jump-go.over-FR.PAST-3PL
'they (PL) jumped over it'

(b) *bubikakaramerandə*
mbu-mbi-kakara-meri-andə
3.ERG-AGAIN-ask-follow-SEQ
'he responded again and'

Incorporated verb constructions expressing sequential events are reasonably common in the narrated texts. Perhaps the most frequent examples are those meaning 'bring' or 'take', which involved the basic motion verbs *e-* ~ *ya-* 'come' or *o-* ~ *wa-* 'go' followed by *ra-* 'stay' followed by *kot-* 'carry', so 'so someone comes or goes for a thing and it remains with them and they then carry it':

(5.98) (a) meŋga   erakotime
            meŋga   e-ra-kot-iya-ome
            who.PC   come-stay-carry-PC-*wh*
            'who (PC) will bring you?'

      (b) mayna   tar o   naɲjen   mbwarakotəkoya
               mayn-na   tar o   naɲjen   mbu-wa-ra-kot-k-oya
               husband-POSS from   child   3.ERG-go-stay-carry-FR.PAST-3SG
               'she took the child from (her) husband'

      (c) nana   ma   gadaerakotənd   ya
               nana   ma   ga-da-e-ra-kot-nd   ya
               mama   1SG   INV-NOW-come-stay-carry-IMP.1.OBJ   EMPH
               'mama, come and get me now!'

      (d) kar   kwarakotəmbiko
               kar   ku-wa-ra-kot-mbi-oko
               feast TR.IMP-go-stay-carry-IM.FUT-2DL
               'you (DL) take the feast!'

Some other examples of incorporated verb constructions with a sequential meaning between the verb roots:

(5.99) (a) ŋgierasakəŋgaya   atik   ŋari
            ŋgi-e-ra-sa-k-ŋgaya   atik   ŋari
            3PL-come-stay-IN-FR.PAST-3PL smoke DAT
            'they (PL) came and stayed inside the smoke'

      (b) mbwarameriananəŋrakududu
               mbu-wa-ra-meri-anəŋra-k-undundu
               3.ERG-go-stay-follow-3PC.DAT-FR.PAST-3PC
               'they (PC) went and followed them (PC)'

      (c) kwarik   abeb   mbutunaɲjakosirikoya
               kwarik   abeb   mbu-t-naɲja-kosiri-k-oya
               side   other   3.ERG-CAUS-be.at.bank-go.across-FR.PAST-3SG
               'he kept to the river bank and then went across to the other side'

(e) *mbu ŋgiteramerik*
   mbu ŋgi-t-e-ra-meri-k
   3PL 3PC-COM-come-stay-follow-FR.PAST
   'they (PC) followed them (PL)'

(f) *karəkarək enamb   karan*
   karəkarək ena-mb   karan
   pillow   PROX.SG-OBL   head
   nandateranəto
   na-nda-t-e-ra-n-to
   1SG.ERG-NOW-CAUS-come-stay-IMP-DEP
   'let me put my head on this pillow and'

(g) *mbiraposirimbaya*
   mbi-rapo-siri-mbaya
   3DL-run-go.down-3DL
   'they (DL) ran and came down'

(h) *mbirasirikəto*
   mbi-ra-siri-k-to
   3DL-stay-go.down-FR.PAST-DEP
   'they (DL) stayed and then went down and'

(i) *tamənd mbutukasayaroronak*
   tamənd mbu-t-kasa-e-ar-oro-ana-k
   fish   3.ERG-CAUS-heap-come-PROG-EXT-3SG.DAT-FR.PAST
   'they were continually heaping up and coming with fish for him'

(j) *kuŋgarəkəratirukorəka*
   ku-ŋga-rək-rati-rukor-ka
   TR.IMP-FIRST-push-pull-go.ashore-2SG
   'you (SG) first push it, pull it so it comes ashore!'

An interesting lexicalized use of the verb root *kata-* 'speak' when incorporated in a serial verb construction is to report hunches or surmises by the speaker:

(5.100) (a) *nəmandəpak mbukatatakwatarandide*
           nəmandəpak   mbu-kata-takwat-ar-andə-nde-e
           woman.PL   3.ERG/1.OBJ-speak-lie-PROG-PFV-1PL-Q
           'I think the women have been lying to us (PL)'

(b) mbam   o   mina   yan   ukataesirande?
    mbam   o   mi-na   yan   u-kata-e-siri-andə-e
    maybe   2SG-POSS papa   3SG-talk-come-go.down-PFV-Q
    'I think maybe your papa has come downriver'

(c) apasen   minak   o   mora   sin   ombe
    apasen   mi-na-k   o   mora   si-n   ombe
    grandparent   2SG-POSS-NE   thing do-NOMZ   one
    ukatakusambəmande
    u-kata-kusa-am-ma-andə-e
    3SG-talk-go.out-DETR-AFTERNOON-PFV-Q
    'I think something has happened to your grandmother during the afternoon'

Chapter 6
Clause structure

6.1 Basic verbal clauses

In Kopar verbal clauses and non-verbal clauses have very different structures and possibilities. Verbal clauses by definition are headed by a verb which defines the number of core arguments, one for an intransitive verb and two for a transitive verb, and their linear order. As a head marking language, Kopar lacks nominal case marking so that the identification of subject and object grammatical functions is accomplished by a mix of verbal agreement and constituent order constraints.

6.1.1 Constituent order in verbal clauses

Kopar like many Papuan languages is essentially verb final. Core arguments like subject and object precede the verb, but oblique arguments marked by postpositions and temporal words can precede or follow, though the order in which they precede (6.1-6.2a) seems favored:

(6.1) (a) normot   rari   nəmbren   nambratondukondu
          normot   rari   nəmbren   nambrat-onduk-ondu
          man.PL   1.day.removed   pig   spear-FUT-3PL
          'the men will spear a pig tomorrow'

     (b) normot   nəmbren   nambratondukondu   rari
          normot   nəmbren   nambrat-onduk-ondu   rari
          man.PL   pig   spear-FUT-3PL   1.day.removed
          'the men will spear a pig tomorrow'

(6.2) (a) Bil   nda   pipikambu   ŋga   ombiya
          Bil   nda   pipikambu   ŋga   o-mbi-oya
          Bill COM   fish.with.hook   DAT   go-IM.FUT-1SG
          'I'm going fishing with Bill'

     (b) pipikambu   ŋga   ombiya   Bil   nda
          pipikambu   ŋga   o-mbi-oya   Bil   nda
          fish.with.hook   DAT   go-IM.FUT-1SG Bill   COM
          'I'm going fishing with Bill'

But names of places, which are not postpositional phrases, follow the verb:

(6.3) (a) *ma   rari   ondukoya   Wewak*
         ma   rari   o-onduk-oya   Wewak
         1SG   1.day.removed   go-FUT-1SG   Wewak
         'I will go to Wewak tomorrow'

   (b) *pipikambu   ŋga   ondək   mandumor*
       pipikambu   ŋga   u-o-ndək   mandumor
       fish.with.hook   DAT   3SG-go-NR.PAST   mangrove
       'he went to fish in the mangroves'

   (c) *mu   urin   mbunambrataŋgoya   Sumbrakambi*
       mu   urin   mbu-nambrat-aŋg-oya   Sumbrakambi
       3SG   crocodile 3.ERG-spear-PRES-3SG   beach
       'he spears a crocodile at the beach'

Kopar verbal clauses in ongoing speech such as the narrated texts normally consist of just a verb and not more than one core argument, either subject or object, more commonly the latter, with an occasional postpositional phrase. Clauses with more than that number of constituents are unusual, so the following sentence, although grammatical, is rather artificial. Still I offer it to exhibit the range of constituent orders in a basic verbal clause that were accepted by speakers as grammatical and those which were not, all provided in response to the prompt 'I picked a coconut for him yesterday'. Note that all are verb final, but allow permutations in the preverbal constituents:

(6.4) (a) *ma   rarindək   muna   ŋga   ruaŋ*
         ma   rari-ndək   mu-na   ŋga   ruaŋ
         1SG   1.day.removed-NR.PAST   3SG-POSS   DAT   coconut
         nasaytundukoya
         na-sayt-ndək-oya
         1SG-pick-NR.PAST-1SG

   (b) ma   rarindək   ruaŋ   muna   ŋga   nasaytundukoya
   (c) rarindək ma   ruaŋ   muna   ŋga   nasaytundukoya
   (d) ma   ruaŋ   rarindək   muna   ŋga   nasaytundukoya
   (e) ?ma   ruaŋ   muna   ŋga   rarindək nasaytundukoya
   (f) *ruaŋ   ma   muna   ŋga   rarindək nasaytundukoya

The sentences are presented in order of decreasing desirability. All speakers agree that (6.4a), with the focus of the immediately preverbal position on *ruaŋ* 'coconut', is the best way to say this sentence, and (6.4f) is decidedly bad. Sentence (6.4f) is bad because of the constituent order is which the object precedes the subject. Speakers in elicitation sessions proscribed this order and overall presented Kopar as a rigid SOV language, though OSV clauses do appear occasionally in the narrative texts. Sentence (6.4e) is a bit odd because *rarindək* 'yesterday' occurs in the immediately preverbal focal position, and it is pragmatically strange to place focus on this meaning when it is already highlighted by the choice of the near past tense, which typically specifies 'yesterday'. Sentences (6.4b-d) are all good, though not the first choice like (6.4a); focus is on the recipient in all three, which simply differ in the placement of the temporal word *rarindək* 'yesterday'.

However, in actual spontaneous speech, such as the narrated texts, proscribed constituent orders, though relatively rare, do occur. For example, heavy noun phrase objects can be extraposed after the verb:

(6.5) (a) *mbusamaytukondu   Paɲjuman   [andi   sambok]*
        mbu-samayt-k-ondu   Paɲjuman   [andi   sambo-k]
        3.ERG-see-FR.PAST-3PL PN   ground   put-FR.PAST
        'they (PL) saw Paɲjuman who was smearing herself with mud'

  (b) *mbusamaytəmak   normot   [tramak]*
        mbu-samayt-ma-k   normot   [tra-ma-k]
        3.ERG-see-DUR-FR.PAST man.PL   dance-DUR-FR.PAST
        'she kept watching the men who were dancing'

As can object NPs which are afterthoughts, though in this case there is a pause and intonation break:

(6.6) (a) *kayn   ŋgitənaɲjak,   timbrəmankayn*
        kayn   ŋgi-t-naɲja-k,   timbrəman-kayn
        canoe   3PL-CAUS-be.at.bank-FR.PAST   sago-canoe
        'they (PL) pulled the canoe to the river bank, a canoe made of sago'

  (b) *nəmand   umbeanakəto   sen,   Wak*
        nəmand   u-mbe-ana-k-to   sen,   Wak
        woman   3SG-bear-3SG.DAT-FR.PAST-DEP   son   PN
        'the woman gave birth to a son, Wak'

Example (6.6b) is interesting in that only the proper name *Wak* is an afterthought and marked by a pause and intonation break. The direct object *sen* 'son' is in the same intonation contour as the verb; an unusual pattern of SVO clausal constituent order for Kopar.

There is only one example in the corpus in which two bare nouns as afterthoughts follow the verb, and it is with the ditransitive verb *kam-* 'give'; even the recipient follows the verb in this case and surprisingly without *ŋga* DAT:

(6.7)   iniŋ   ŋgitonak,   ŋgikamanak,
       iniŋ   ŋgi-t-o-na-k,   ŋgi-kam-ana-k,
       stone 3PC-CAUS-go-UP-FR.PAST 3PC-give-3SG.DAT-FR.PAST
       Wak,   iniŋ   yo
       Wak,   iniŋ   yo
       PN   stone DEF
       'they (PC) lifted up a stone and gave it to him, to Wak, the stone'

6.1.2 The expression of grammatical functions in verbal clauses

6.1.2.1 Grammatical relations in intransitive and transitive clauses

As befits its head marking typology, Kopar lacks any case marking for the core arguments of verbs, subjects of intransitive verbs and subjects and objects of transitive verbs. Noun phrases in the elication sessions were often followed by a particle *o*, though this occurred more rarely in the narrated texts. This particle *o* is not restricted to any particular grammatical function, so cannot be analyzed as a case marker:

(6.8) (a)   nor   o   təndasana
           nor   o   t-Ø-ndasa-n-a
           man   PFV-3SG-sit-LK-PFV
           'the man sat down'

      (b)   mbu indan   o   mbundimanagodu
           mbu indan   o   mbu-undi-ma-n-aŋg-ondu
           3PL   house   3.ERG-build-DUR-LK-PRES-3PL
           'they are building a house'

      (c)   rari   tumbuna   porakay o
           rari   tumbuna   porakay o
           1.day.removed morning   path
           makarosirindəkənaya
           ma-kar-o-siri-ndək-naya
           1SG-walk-go-go.down-MORNING-NR.PAST-1SG
           'I walked down along the path yesterday morning'

(d) *ma   kay   sur   o   təmarəma*
    ma   kay   sur   o   t-ma-rəmə-a
    1SG   canoe inside   PFV-1SG-stand-PFV
    'I've stood up inside the canoe'

In (6.8a) *o* occurs with the subject, in (6.8b) with the object, in (6.8c) with an oblique without a postposition, and in (6.8d) with an oblique which is a postpositional phrase. Clearly *o* cannot be a case marker and seems more likely to be a prosodic or pragmatic feature, although one whose function remains unknown.

The grammatical relations of subject and object are signaled by a combination of verb agreement and word order in Kopar. The complex system of verb agreement by bound pronominals was discussed in depth in section 5.2. Baring the presence of a dative pronominal agreement suffix for the various functions discussed in section 5.2.4, transitive verbs typically only agree with one core argument in Kopar, either subject or object depending on the Animacy Hierarchy, so verb agreement is commonly insufficient to distinguish subjects from objects, hence a fairly strict constituent order of subject before object is commonly mandated in the language, though the presence of overt subjects and objects in a single clause is itself a rare and marked structure. Normally, given the one core argument constraint per clause that generally holds in the language, they would be spread over two clauses. However, in the rare and somewhat artificial situation of an overt subject and object in the same clause, even when animacy constraints should be sufficient to determine grammatical functions regardless of the depauperate verbal morphology, speakers in elicitation prefer the order SOV:

(6.9) (a) *nor   indan   mbundiaragodu*
          nor   indan   mbu-undi-ar-aŋg-ondu
          man   house   3.ERG-build-PROG-PRES-3PL
          'the men are building a house'

      (b) *???indan   nor   mbundiaragodu*
          indan   nor   mbu-undi-ar-aŋg-ondu
          house   man   3.ERG-build-PROG-PRES-3PL
          'the men are building a house'

Nor does the ranking of person, so relevant in the system of bound pronominals, have any role in linear order of constituents; subjects precede objects regardless of person:

(6.10) (a) *mu*   *ma*   *ŋasamaytundəkənaya*
            mu   ma   ŋa-samayt-ndək-naya
            3SG   1SG   INV-see-NR.PAST-1SG
            'she saw me'

     (b) **ma*   *mu*   *ŋasamaytəndukənaya*
            ma   mu   ŋa-samayt-ndək-naya
            1SG   3SG   INV-see-NR.PAST-1SG
            'she saw me'

     (c) *ma*   *nor*   *nasamaytaŋgoya*
            ma   nor   na-samayt-aŋg-oya
            1SG   man   1SG.ERG-see-PRES-3
            'I see the man'

     (d) **nor*   *ma*   *nasamaytaŋgoya*
            nor   ma   na-samayt-aŋg-oya
            man   1SG   1SG.ERG-see-PRES-3
            'I see the man'

The linear order of (6.10d) is only possible if *nor* 'man' is taken as the subject, and then, of course, a verb with inverse inflection is required:

(6.11) *nor*   *ma*   *ŋasamaytaŋgənaya*
        nor   ma   ŋa-samayt-aŋg-naya
        man   1SG   INV-see-PRES-1SG
        'the man sees me'

although a textual example of exactly the rejected order of (6.10d) is found in (3.27g).

6.1.2.2 Verbs with cognate objects
Verbs with cognate objects are formally intransitive in Kopar, though some like *mə-* 'eat' are ambitransitive:

(6.12) (a) *nor*   *məndə*   *upraraŋgoya*
            nor   məndə   u-pra-ar-aŋg-oya
            man   feces   3SG-excrete-PROG-PRES-3SG
            'the man is defecating'

(b) *mumoran təmama*
 mə-moran t-ma-mə-a
 eat-thing PFV-1SG-eat-PFV
 'I've eaten'

(c) *sara ukataraŋoya*
 sara u-kata-ar-aŋ-oya
 report 3SG-speak-PROG-PRES-3SG
 'he is reporting something'

(d) *nor pwar upriaraŋoya*
 nor pwar u-pri-ar-aŋ-oya
 man song 3SG-sing-PROG-PRES-3SG
 'the man is singing'

This also applies to idioms with adjuncts. If the verb root is intransitive, it is inflected intransitively, as the adjunct does not count as a core argument:

(6.13) *paŋgə arəm ikirindikiya*
 paŋgə arəm i-ki-ri-ndək-iya
 1PC water 1-bathe-DOWN-NR.PAST-PC
 'we bathed yesterday'

6.1.2.3 Dative arguments and ditransitive clauses

As apparent from a number of examples in this monograph, ditransitive verbs present a complex picture in Kopar. While there is a set of dative pronominal suffixes which can indicate the recipient of a ditransitive verb, because of the Animacy Hierarchy and the direct-inverse system that operates in the language, this is restricted only to non-local third person recipients. A similar restriction on dative affixes operates in its sister language Yimas (Foley 1991:208–216) and indeed the quite unrelated Romance languages. Consider the prototypical canonical ditransitive verb *kam-* 'give'. Example (6.14a) is grammatical and the usual structure, but example (6.14b) is decidedly ungrammatical:

(6.14) (a) *iniŋ mbukamanakoya*
 iniŋ mbu-kam-ana-k-oya
 stone 3.ERG-give-3SG.DAT-FR.PAST-3SG
 'he gave him a stone'

(b) *iniŋ mbukamanaŋgakoya
    iniŋ mbu-kam-anaŋga-k-oya
    stone 3ERG-give-1SG.DAT-FR.PAST-3SG
    'he gave me a stone'

The closest equivalent to the intended meaning of (6.14b) is to use an inverse structure, where the recipient agrees with suffixal pronominal agreement, but this is uncommon:

(6.15) iniŋ ŋakaməkənaya
       iniŋ ŋa-kam-k-naya
       stone INV-give-FR.PAST-1SG
       'he gave me a stone'

And if both the subject and the recipient are local persons, not even this is possible; in this case the verb reverts to being monotransitive, according to the pattern of a transitive verb with a second person subject acting on a third person object or a first person subject acting on a third person object. The local recipient is simply marked with the dative postposition ŋga:

(6.16) (a) mana   ŋga   iniŋ   ikamaŋona
           ma-na   ŋga   iniŋ   i-kam-aŋg-ona
           1SG-POSS DAT   stone   2-give-LK-PRES-2SG
           'you (SG) give me a stone'

       (b) mina   ŋga   iniŋ   ikamaŋokə
           mi-na   ŋga   iniŋ   i-kam-aŋg-okə
           2SG-POSS DAT   betelnut 1-give-PRES-1PC
           'we (PC) give you (SG) a stone'

(For some minor variations on the pattern in (6.16), see section 5.2.2.5; though the verbs in these configurations always inflect monotransitively).

Because of the complexities presented by the restrictions on inflectional possibilities, the expression of the meaning 'give' in Kopar is itself quite interesting. Besides the canonical verb root *kam-* 'give', there are two other derived verb stems that correspond to this meaning, and both are derived from intransitive verb roots plus the causative prefix *t-*, and by this derivation can only have two core arguments, i. e. are monotransitive and so simplify the difficulties presented above. One is native *t-e-* CAUS-come 'bring, give' and another almost certainly involves a Tok Pisin borrowing *t-kam* CAUS-arrive 'give'. The use of these contrasting verbs now seems specialized to the person of the recipient, the former for first person recipients and the latter for second and third person recipients. However, the former can also be used when the deictic center, for example in a narrative text, is a non-local third person, in which case the recipient can be indicated by a dative suffix, although here the meaning is always 'bring' (6.17c):

(6.17) (a) nor   mana   ŋga   uren   mbutenaŋgoya
           nor   ma-na   ŋga   uren   mbu-t-e-n-aŋg-oya
           man   1SG-POSS   DAT   dog   3.ERG-CAUS-come-LK-PRES-3SG
           'the man gives me a dog'

       (b) ma   nor   ŋga   uren   natəkamaŋgoya
           ma   nor   ŋga   uren   na-t-kam-aŋg-oya
           1SG   man   DAT   dog   1SG.ERG-CAUS-arrive-PRES-1SG
           'I give the man a dog'

       (c) nana   nəmbre   okumbi   mbutetanak
           nana   nəmbre   okumbi   mbu-t-e-ta-ana-k
           mama   pig   PL   3.ERG-CAUS-come-OUT-3SG.DAT-FR.PAST
           'mama brought pigs out for him'

This relative restriction to first person recipients, as in (6.17a) and (6.18), of *t-e-* CAUS-come 'give' restrictions makes sense in the light of a speech act deictic meaning of 'toward speaker' of *e-* 'come':

(6.18) (a) mbə   kena   ŋga   sokayn   mbitenaŋgodə
           mbə   ke-na   ŋga   sokayn   mbi-t-e-n-aŋg-ondə
           3DL   2DL-POSS   DAT   tobacco   3DL-CAUS-come-LK-PRES-3DL
           'they (DL) give us (DL) tobacco'

       (b) numotanda   ena   ŋga   puruŋ   mbutendukoya
           numotanda   e-na   ŋga   puruŋ   mbu-t-e-ndək-oya
           man.PL   1PL-POSS   DAT   betelnut   3.ERG-CAUS-come-NR.PAST-3
           'the people gave us (PL) betelnut'

In fact, *t-e-* CAUS-come can contrast with *t-o* CAUS-go on exactly this deictic dimension:

(6.19) (a) *mana   ŋga   awr   kuteka*
          ma-na   ŋga   awr   ku-t-e-ka
          1SG-POSS   DAT   fire TR.IMP-CAUS-come-2SG
          mumora   si   ŋga
          mə-mora   si   ŋga
          eat-thing   make   DAT
          'bring me fire to cook food!'

(b) *muna   ŋga   awr   kutoka   mumora   si   ŋga*
    mu-na   ŋga   awr   ku-t-o-ka   mə-mora   si   ŋga
    3SG-POSS   DAT   fire TR.IMP-CAUS-go-2SG eat-thing make   DAT
    'bring him/her fire to cook food!

The other form for 'give' is used with second and third person recipients. Its root is *kam-*, homophonous with the native root *kam-* 'give', but almost certainly a loan from Tok Pisin *kamap* 'arrive, become', the loss of the final *ap* syllable also being found in another loan, *bagari-* 'kill' from Tok Pisin *bagarap* 'ruin, destroy, harm someone'. Further evidence that is is indeed a borrowing from Tok Pisin is the fact that as in Tok Pisin the same verb root is used to mean 'become', often usurping the native verb root *si-* 'do, make, become, happen, feel' for that meaning. To this borrowed root *kam-* 'arrive' is added the causative prefix *t-*, so that 'give' here is transparently 'cause to arrive', i. e. cause something to arrive at someone', that someone being a second or third person recipient, as in (6.17b) and (6.20):

(6.20) (a) *ma   mbuna   ŋga   puruŋ   natəkamundukoya*
          ma   mbu-na   ŋga   puruŋ   na-t-kam-ndək-oya
          1SG   3PL-POSS   DAT   betelnut   1SG.ERG-CAUS-arrive-NR.PAST-1SG
          'I gave them (PL) betelnut'

(b) *paŋgə   kona   ŋga   sokay   itəkamaŋgok*
    paŋgə   ko-na   ŋga   sokay   i-t-kam-aŋg-okə
    1PC   2DL-POSS   DAT   tobacco   1-CAUS-arrive-PRES-1PC
    'we (PC) give you (DL) tobacco'

In spite of the encoding of the person of the recipient in the choice of the verb stem *t-e-* CAUS-come 'give.1' versus *t-kam* CAUS-arrive 'give.2/3', local recipients, because they are prohibited from occurring as dative suffixes in ditransitive constructions, are not generally treated as core arguments of the verb. Evidence for this comes from erstwhile inverse constructions in which the recipient is a higher ranked person than the giver. Instead of the inverse inflection that we would expect if the recipient was the direct object of the verb, we typically get neutral inflection, that is a non-local third person giver acting on a non-local third person transferred object, with that transferred object functioning as the direct object grammatical relation and the recipient as an oblique:

(6.21) məŋgə mana ŋga uren ŋgitenaŋgoya
məŋgə ma-na ŋga uren ŋgi-t-e-n-aŋg-oya
3PC 1SG-POSS DAT dog 3PC-CAUS-come-LK-PRES-3SG
'they (PC) give me a dog'

This what we usually find, but there are exceptional inverse cases indicating the status of a local person recipient with or without oblique marking with ŋga DAT. Consider this example:

(6.22) məŋgə ma dəbir ŋgakariaŋgaya
məŋgə ma dəbir ŋga-kari-aŋg-aya
3PC 1SG regalia INV-put-PRES-1SG
'they (PC) put ceremonial regalia/bling (Tok Pisin *bilas*) on me'

Note the lack of the dative postposition for the recipient in (6.22) and agreement for it with the suffix –*aya* 1SG, with the verb marked as inverse. In (6.22) all three arguments are bare and hence treated as core arguments, a transparently ditransitive clause, but such examples are rare in the corpus.

Speakers also now occasionally treat *t-kam-* CAUS-arrive in the same say as the canonical verb *kam-* 'give', not surprisingly given their shared meaning, so we sometimes meet with sentences in which *t-kam* CAUS-arrive takes non-local dative suffixes to indicate the recipient:

(6.23) muɲamb mbutukamanak
muɲamb mbu-t-kam-ana-k
sugarcane 3.ERG-CAUS-arrive-3SG.DAT-FR.PAST
'he gave her sugarcane'

It is important to remember that the mere presence of a dative pronominal agreement suffix does not make a verb ditransitive for inflectional purposes, even if it indicates a recipient participant. If the valence of the verb stem is intransitive, it remains intransitive even with a dative suffix:

(6.24) (a) munəmb yowa ukusamanəmbakoya
munəmb yowa u-kusa-am-anəmba-k-oya
smell DIST 3SG-go.out-DETR-3DL.DAT-FR.PAST-3SG
'that smell went over to them (DL)'

(b) sara   ukatanumbwakoya
    sara   u-kata-anumbwa-k-oya
    story 3SG-speak-3PL.DAT-FR.PAST-3SG
    'he told them (PL) the story'

Note that both of these verbs inflect intransitively with *u-* 3SG instead of *mbu-* 3.ERG. Both verb roots are inherently intransitive, *kusa-* 'go out' rather obviously semantically (and it also takes *–am* DETR here), but *kata-* 'speak' because it is a cognate object verb root and as pointed out in section 6.1.2.2, such verb roots are inherently intransitive.

6.1.2.4 Experiential clauses

Experiential clauses are those that express meanings like 'I feel sick' or 'I am hungry'. Note that in English the subject of these sentences is the experiencer of the feeling and the cause or stimulus of the feeling is expressed as the predicate. Kopar expresses such meanings somewhat in a reverse way to Modern English. The verb is impersonal, rather like *methinks* in Middle English. The stimulus or cause of the experience can be unmarked (6.25a, b) or indicated by the dative postposition *ŋga* (6.25d, e). The subject pronominal agreement can only be impersonal, always *u-* 3SG, while the experiencer, if marked by verbal agreement is expressed by a dative pronominal agreement suffix. The verb used in these constructions is the generic verb *si-* 'do, make, happen, become, feel':

(6.25) (a) *kandəknambrin usinaŋgak*
      kandək-nambrin u-si-ananga-k
      sleep-eye   3SG-do-1SG.DAT-FR.PAST
      'I felt sleepy'

   (b) *karo   kaynda   imbotma usiaroronakoya*
      kar-o   kay-onda imbotma u-si-ar-oro-ana-k-oya
      walk-go NEG-3SG jealousy 3SG-do-PROG-EXT-3SG.DAT-FR.PAST-3SG
      'he didn't walk around, he continually felt jealous'
      (literally 'jealousy did to him')

   (c) *nəmand yowa surun nda usianak*
      nəmand yowa surun nda u-si-ana-k
      woman   DIST   belly   COM 3SG-do-3SG.DAT-FR.PAST
      'she was pregnant' (this is likely a calque of Tok Pisin *i gat bel* 'to be pregnant')

(d) miɲjir   pra   ŋga   usianakoya
    miɲjir   pra   ŋga   u-si-ana-k-oya
    urine   excrete DAT   3SG-do-3SG.DAT-FR.PAST-3SG
    'she felt like urinating'

(e) ŋgiramakəto   nda   wakəna
    ŋgi-ra-ma-k-to   nda   wakəna
    3PL-stay-AFTERNOON-FR.PAST-DEP   and   afternoon
    nana   e   ŋga   usianakoya
    nana   e   ŋga   u-si-ana-k-oya
    mama   come DAT   3SG-do-3SG.DAT-FR.PAST-3SG
    'they stayed until afternoon and in the afternoon mama wanted to come'

Although rare, it is possible to express the experiencer as subject, though this again is probably a calque from Tok Pisin:

(6.26) ma   mora   kanda   masimanaŋgənaya
        ma   mora   kanda   ma-si-ma-n-aŋg-naya
        1SG   thing   sick   1SG-feel-DUR-LK-PRES-1SG
        'I feel sick'

And the dative suffixes are not attested with perfective aspect inflection:

(6.27) (a) ma   iɲarapin   təsya
            ma   iɲarapin   t-Ø-si-a
            1SG   shame   PFV-3SG-do-PFV
            'I felt ashamed'

       (b) ma   nime   surun   təsya
            ma   nime   surun   t-Ø-si-a
            1SG   hunger   belly   PFV-3SG-do-PFV
            'I feel hungry'

6.1.3 Marking of oblique roles by postpositions

6.1.3.1 The dative postposition *ŋga*

Besides marking the recipient role with verbs of giving discussed in section 6.1.2.2, the dative postposition *ŋga* has a wide range of functions associated with dative case in other languages, such as persons being benefited (6.28a, b, c) or harmed (6.28d, e) by an action, or persons desired or things attended to by an action (6.29). It is also used with verbs in interclausal relations to indicate infinitives (see section 7.1.1), rather as English *to* covers a similar range. Pronominal complements of *ŋga* DAT always co-occur with the possessive suffix *-na*. Here are some examples of its benefactive or malefactive use:

(6.28) (a) *muna   ŋga   sokayn   naukindukoya*
           mu-na   ŋga   sokayn   na-uki-ndək-oya
           3SG-POSS   DAT   tobacco   1SG.ERG-buy-NR.PAST-1SG
           'I bought tobacco for him'

   (b) *o   mobido   kumuobido*
       o   ma-o-mbi-ondo   ku-mə-o-mbi-ondo
       2PL   ITR.IMP-go-IM.FUT-2PL TR.IMP-eat-go-IM.FUT-2PL
       porakay   yo,   mana   ŋga   awr   sur
       porakay   yo,   ma-na   ŋga   awr   sur
       path   DEF   1SG-POSS   DAT   fire   in
       'you (PL) go, eat as you along along the path, (leave the food) for me in the fire'

   (c) *sen   ŋga   ipunumanaŋgəbaya*
       sen   ŋga   i-punu-ma-n-aŋg-mbaya
       son   DAT   1-work.sago-AFTERNOON-LK-PRES-DL
       'we (DL) are working sago for the son in the afternoon'

   (d) *nambwe   ŋga   numotanda   naŋgrin   ŋgerakotək*
       nambwe   ŋga   numotanda   naŋgrin   ŋgi-e-ra-kot-k
       PROX.PL   DAT   man.PL   spear   3PL-come-stay-carry-FR.PAST
       'for these (PL), the men brought spears' (intending to kill them)

   (e) *mbəna   ŋga   awrporakay   mbukarikududu*
       mbə-na   ŋga   awr-porakay mbu-kari-k-undundu
       3DL-POSS   DAT   fire-path   3.ERG-put-FR.PAST-3PC
       'they (PC) went on the war path against them (DL)' ('fire-path' is the idiom for a state of hostilities between parties)

Here are examples of its use with a person or thing which are the objects of desire or attention:

(6.29) (a) *ma   nunon   kapuya   mana   nəma   ŋga*
           ma   nunon   kapu-oya   ma-na   nəma   ŋga
           1SG   desire   big-1SG   1SG-POSS   woman   DAT
           'I have a big desire for my wife'

(b) *mi   sokayn   ŋga   rine*
    mi   sokayn   ŋga   ri-ona-e
    2SG tobacco DAT   want-2SG-Q
    'do you (SG) want cigarettes?'

(c) *kunden   ŋga   moka*
    kunden   ŋga   ma-o-ka
    black.palm DAT ITR.IMP-go-2SG
    'you (SG) go for black palm!'

(d) *Wak   iniŋ   ŋga   urak*
    Wak   iniŋ   ŋga   u-ra-k
    PN   stone DAT   3SG-stay-FR.PAST
    'Wak stayed for the stone' (in order to get it)

(e) *ma   tamənd   ŋga   masak   sur   nambrin*
    ma   tamənd   ŋga   masak   sur   nambrin
    1SG fish   D   ocean   inside eye
    natərəmaraŋoya
    na-t-rəmə-ar-aŋg-oya
    1SG.ERG-CAUS-stand-PROG-PRES-1SG
    'I'm fixing my eye on the fish in the ocean'

(f) *mbu   jadək   ŋga   ŋgiok*
    mbu   jadək   ŋga   ŋgi-o-k
    3PL   nothing DAT   3PL-go-FR.PAST
    'they (PL) went for nothing' (got no result)

(g) *puruŋ   ŋga   napar   mbutukaməmanak*
    puruŋ   ŋga   napar   mbu-t-kam-ma-ana-k
    betelnut DAT hand   3.ERG-CAUS-arrive-DUR-3SG.DAT-FR.PAST
    'he kept putting out his hand for betelnut'

The word *nunon* in (6.29a) covers the range of English 'mind, thought, feelings, desires', essentially the conscious state of thoughts, feelings and beliefs. In (6.29e) the phrase *nambrin t-rəmə-* eye CAUS-stand is essentially the Kopar equivalent of English 'got my eye on'; an alternative idiom with similar meaning is *nambrin t-kam* eye CAUS-arrive.

There is a variant of *ŋga* DAT of the form *ŋgari(n)*, which has arisen from a re-analysis and grammaticalization of the sequence of *ŋga* and the desiderative verb *ri-* described in section 7.1.1. It is used to denote a strong wish on the part of the actor that a goal be reached (6.30a, b) or that a recipient get an object (6.30c):

(6.30) (a) ŋgierasakəngaya   atik   ŋgari
           ngi-e-ra-sa-k-ngaya   atik   ngari
           3PL-come-stay-IN-FR.PAST-3PL smoke   DAT
           'they (PL) came and stayed inside the smoke'

   (b) kutururujabiduku   paret   ŋgari
       ku-t-ruruɲja-mbi-onduku   paret   ngari
       TR.IMP-CAUS-shake-IM.FUT-2PC outside DAT
       'you (PC) shake (it) until (it comes) out!'

   (c) kakandək   ŋgari sora   kamək
       kakandək   ngari sora   kam-k
       older.sister DAT   shell.type give-FR.PAST
       nda ŋgiramak
       nda ngi-ra-ma-k
       and 3PC-stay-AFTERNOON-FR.PAST
       ŋgipunumak
       ngi-punu-ma-k
       3PC-work.sago-AFTERNOON-FR.PAST
       'she gave the older sister a shell and they (PC) stayed until afternoon and worked sago'

6.1.3.2 The purposive postposition *ndək*

It remains unclear whether this morpheme should be analyzed as a suffix or postposition, as the data are equivocal. One of its meanings of is quite close to that of *ŋga* DAT, and sometimes they are interchangeable (6.31b). Its core meaning is also 'for' when the purpose of the action is the acquisition or achievement of something. Like *ŋga* DAT it is also used with verbs to indicate purposive infinitives, and this is perhaps its more common usage (see section 7.1.2). But unlike *ŋga* DAT, it never co-occurs with -*na* POSS and further can be bound to its complement as a single phonological word because it undergoes denasalization following a prenasalized voiced stop (6.31a), which *ŋga* DAT never does:

(6.31) (a) naŋgep   indadək   nakotaraŋgoya
           nangep   inda-ndək   na-kot-ar-ang-oya
           house.post   house-PURP   1SG-carry-PROG-PRES-1SG
           'I'm carrying posts for a house'

   (b) mumoran   muna   ŋga
       mə-moran   mu-na   nga
       eat-thing 3SG-POSS DAT

	nasiarundukoya		tra	ndək/ŋga
	na-si-ar-ndək-oya		tra	ndək/ŋga
	1SG.ERG-make-PROG-NR.PAST-1SG		dance	PURP/DAT

'I made food for him/her for the dance'

Interestingly, *ndək* PURP can also translate English 'from' if the purpose of the action is the removal of something; in this usage it co-occurs with the particle *o*:

(6.32) (a) *kambowen asumb o ndək ŋgiparianak*
kambowen asumb o ndək ŋgi-pari-ana-k
spines mouth PURP 3PL-extract-3SG.DAT-FR.PAST
'they removed the spines from his mouth'

(b) *nimbep napar sur o ndək urotanakəto*
nimbep napar sur o ndək u-rot-ana-k-to
spear hand inside PURP 3SG-fall-3SG.DAT-FR.PAST-DEP
'the spear fell from his hand and'

(c) *rakondək eranakə, karanpi ŋari*
rakon-ndək e-ra-ana-k karan-pi ŋari
shoulder-PURP come-stay-3SG.DAT-FR.PAST head-just DAT
'coming from her shoulder, just to her head'

Note the presence of the prosodic particle *o* in (6.32) blocks the binding of the postposition to its noun complement and prevents denasalization in (6.32a); it is not known if *o* is required when the postposition has this source function (example (6.32c) is not clear counterevidence in that *rakondək* could underlyingly be *rako o ndək*, with missing final /n/ common for nouns and the prosodic particle truncated because of the identical preceding vowel).

6.1.3.3 The source postposition *ta(r)* ~ *tar o ndək*

As with *sur* 'in, into, inside, on', the final /r/ of *tar-* SOURCE is often omitted. It appears these two forms can be used interchangeably, and no meaning difference was noted. The longer form appears to be derived from the shorter plus the addition of the prosodic particle *o* and the purposive postposition; essentially this is the source usage of *-ndək* PURP exemplified in (6.32) in combination with *tar* SOURCE. Only a few examples of this were attested in the corpus:

(6.33) (a) ma   muna   tar o ndək ruaŋ   naerakotundukoya
        ma   mu-na   tar o ndək ruaŋ   na-e-ra-kot-ndək-oya
        1SG   3SG-POSS   SOURCE   coconut   1SG-come-stay-carry-NR.PAST-1SG
        'I brought a coconut from her'

   (b) umbikrarik   mbu   tar   ambiri
       u-mbi-krari-k   mbu   tar   ambiri
       3SG-AGAIN-hide-FR.PAST   3PL   SOURCE   alone
       pamba   ŋgiwarorok
       pamba   ŋgi-o-ar-oro-k
       only   3PL-go-PROG-EXT-FR.PAST
       'he hid from them (PL) again, and they (PL) were going (away) by themselves for a while'

   (c) mayna   tar o   naɲjen   mbwarakotukoya
       mayn-na   tar o   naɲjen   mbu-wa-ra-kot-k-oya
       husband-POSS SOURCE   child   3.ERG-go-stay-carry-FR.PAST-3SG
       'she took the child from (her) husband'

   (d) Kotokari   yo   urikeranak,
       Kotokari   yo   u-riker-ana-k
       PN   DEF   3SG-get.up-3SG.DAT-FR.PAST
       ekəto   apasen   tar o
       e-k-to   apasen   tar o
       come-FR.PAST-DEP   grandparent   SOURCE
       'Kotokari rose up from him, leaving his grandparent (behind)'

6.1.3.4 Locative postpositions

Other than the use of the oblique case marker discussed in section 6.1.3.6, no general locative postposition has been identified for Kopar. Names of places, whether purely locational or directional, are left unmarked and commonly follow the verb:

(6.34) (a) rari   tumbuna   porakay
           rari   tumbuna   porakay
           1.day.removed   morning   path
           makarosirindəkənaya
           ma-kar-o-siri-ndək-naya
           1SG-walk-go-go.down-MORNING-NR.PAST-1SG
           'I walked down the path yesterday morning'

(b) *paŋə mandaokə   Sumbrakambi*
    paŋə ma-nda-o-okə   Sumbrakambi
    1PC   ITR.IMP-NOW-go-1PC beach
    'let us (PC) go to the beach right now'

(c) *mu   urin   mbunambrataŋoya   Sumbrakambi*
    mu   urin   mbu-nambrat-aŋ-oya Sumbrakambi
    3SG   crocodile 3.ERG-spear-PRES-3SG beach
    'he spears a crocodile at the beach'

(d) *ma   rari   ondukoya   Wewak*
    ma   rari   o-onduk-oya Wewak
    1SG   1.day.removed   go-FUT-1SG   Wewak
    'I will go to Wewak tomorrow'

(e) *pipikambu   ŋga ondək   mandumor*
    pipikambu   ŋga o-ndək   mandumor
    fish.with.hook DAT go-NR.PAST mangrove
    'she went to fish in the mangroves'

Otherwise, the locational postpositions are used, most commonly *sur* 'in, inside, into, on'; these postpositional phrases most commonly precede the verb, but following is also possible (6.35f):

(6.35) (a) *moran timbrak   kaŋgarap   mbukariaŋgududu*
           moran   timbrak   kaŋgarap   mbu-kari-aŋg-undundu
           thing   platform on.top.of 3.ERG-put-PRES-3PC
           'they (PC) put food on top of the platform'

(b) *nor   iror   kaŋgarap   təndasana*
    nor   iror   kaŋgarap   t-Ø-ndasa-n-a
    man   tree   above   PFV-3SG-sit-LK-PFV
    'the man has sat on top of the tree/wood'

(c) *tamənd   nani   sur   nakariaŋgoya*
    tamənd   nani   sur   na-kari-aŋg-oya
    fish   pot   inside 1SG.ERG-put-PRES-1SG
    'I put the fish into the pot'

(d) *ma   kay   sur   təmarəma*
    ma   kay   sur   t-ma-rəmə-a
    1SG   canoe   inside PFV-1SG-stand-PFV
    'I've stood up in the canoe'

(e) tamənd   masak   sur   utapwataŋoya
    tamənd   masak   sur   u-tapwat-aŋg-oya
    fish   ocean   inside   3SG-swim-PRES-3SG
    'fish swim in the ocean'

(f) mbutotakududu   numot   sur
    mbu-t-o-ta-k-undundu   numot   sur
    3.ERG-CAUS-go-OUT-FR.PAST-3PC   village inside
    'they (PC) brought them out (from where they
    were) into the village'

(g) nor   inda   sur   ukandəkseaŋoya
    nor   inda   sur   u-kandək-se-aŋg-oya
    man   house   inside   3SG-sleep-NIGHT-PRES-3SG
    'the man sleeps inside the house'

The postposition *sur* 'in, inside, into, on' is interesting as it is very flexible in its categoriality. Besides being a postposition, it functions as the base of a noun *suru(n)* 'belly, stomach', as well as a verb root:

(6.36) (a) asumb   sur   mbusurorarianakoya
           asumb   sur   mbu-sur-o-ra-ri-ana-k-oya
           mouth   in   3.ERG-inside-go-stay-DOWN-3SG.DAT-FR.PAST-3SG
           'he jammed it down into her mouth'
       (b) aweramb   sur   o   mbusurerarik
           aweramb   sur   o   mbu-sur-e-ra-ri-k
           mosquito.net   in   3.ERG-inside-come-stay-DOWN-FR.PAST
           'she put it down inside the mosquito net'

6.1.3.5 The comitative postposition *nda*

As with *ŋga* DAT and *ta(r)* SOURCE, this postposition always co-occurs with *-na* POSS if its complement is pronominal. This postposition is homophonous with the conjunction *nda* 'and', but can be distinguished from it in that the latter does not co-occur with *-na* POSS. Like *ndək* PURP, this is sometimes bound as a suffix and undergoes denasalization. In the elicitation sessions, this postposition was easily obtained, but in the narrative texts it is less common; speakers prefer to use conjunction with *nda* 'and' or the comitative applicative *t-* (section 5.4.2.2.1):

(6.37) (a) *Pet   nda   pipikambu   ŋga   obida*
Peta   nda   pipikambu   ŋga   o-mbi-onda
Peter COM   fish.with.hook DAT   go-IM.FUT-3SG
'he is going to go fishing with Peter'

(b) *Keli   nda   sara   kata   ŋga   riya*
Keli   nda   sara   kata   ŋga   ri-oya
Kelly COM   report speak   DAT   want-1SG
'I want to speak with Kelly'

(c) *ma   mbəna   nda   indan   naundindukoya*
ma   mbə-na   nda   indan   na-undi-ndək-oya
1SG   2DL-POSS   COM   house   1SG.ERG-build-NR.PAST-1SG
'I built a house with them (DL)'

(d) *ŋgeroŋ   nda   utrak*
ŋgeroŋ   nda   u-tra-k
happiness COM   3SG-dance-FR.PAST
'he danced with joy'

(e) *nəmad   yowa   surun   nda   usianak*
nəmad   yowa   suru   nda   u-si-ana-k
woman   DIST   belly   COM   3SG-feel-3SG.DAT-FR.PAST
'she was pregnant' (this is a calque of Tok Pisin *i gat bel* 'to be pregnant')

6.1.3.6 The oblique case marker *-mb*

This suffix is most commonly employed with verbs to indicate adverbial clauses (see section 7.2.1), but can also be used with nouns to mark a number of semantic notions. First, it is used to indicate instruments:

(6.38) (a) *nor   nəmbren   inamasemb   mbusiranaŋgoya*
nor   nəmbren   inamase-mb   mbu-sira-n-aŋg-oya
man   pig   knife-OBL   3.ERG-cut.up-LK-PRES-3SG
'the man is cutting the pig up with a knife'

(b) *Wak   ŋgara   iniŋ   enamb   Sapend*
Wak   ŋgara   iniŋ   ena-mb   Sapend
PN   in.front   stone   PROX.SG-OBL   PN
mbukwayraməkondu
mbu-kwayram-k-ondu
3.ERG-break.up-FR.PAST-3PL
'Wak in front, they (PL) smashed Sapend up with this stone'

(c) karəkarək   enamb   karan
    karəkarək   ena-mb   karan
    pillow   PROX.SG-OBL   head
    nandateranəto
    na-nda-t-e-ra-n-to
    1SG.ERG-NOW-CAUS-come-stay-IMP-DEP
    'let me put my head on this pillow now and'

(d) pwar   enamb   maprimbiko
    pwar   ena-mb   ma-pri-mbi-oko
    song   PROX.SG-OBL   ITR.IMP-sing-IM.FUT-2DL
    'you (DL) should sing this song'

Example (6.38c) perhaps has as much of a locative meaning as an instrumental one.

This suffix also marks the cause or reason for some action; instrumental and causal is a very common polysemy in the languages of the world:

(6.39) (a) muna   nuno   enamb   naokəto
           mu-na   nuno   ena-mb   na-o-k-to
           3SG-POSS   thought   PROX.SG-OBL   3SG-go-FR.PAST-DEP
           'he went because of this thought of his and'

       (b) enamb   usaranəŋgrakoya
           ena-mb   u-sara-nəŋgra-k-oya
           PROX.SG-OBL   3SG-report-3PC.DAT-FR.PAST
           'he told them (PC) about this'

This suffix also sometimes has a clear general locative use (see also 6.38c):

(6.40) (a) ma   numot   ena-mb   mararaŋgaya
           ma   numot   ena-mb   ma-ra-ar-aŋg-aya
           1SG   village PROX.SG-OBL   1SG-stay-PROG-PRES-1SG
           'I am staying in this village'

       (b) paŋgə   inda   aratəkəmb
           paŋgə   inda   aratək-mb
           1PC   house   good-OBL
           təmbitenandəkəkə
           t-mbi-t-e-n-andə-kəkə
           PFV-3.ERG=2/1.OBJ-CAUS-come-LK-PFV-1PC
           'you have brought us (PC) to a good house'

Note impersonalization here, with a second person subject acting on a first person object, with one of the variants of *mbu-* 3.ERG discussed in section 5.2.2.3.

    (c)   *mana   inda   sur sambot   ŋgari*
         ma-na   inda   sur sambo-t   ŋgari
         1SG-POSS house in   leave-APPL DAT
         mana   irikanəmb
         ma-na   irikan-mb
         1SG-POSS front.of.house-OBL
         'to leave him in my house, at the front of my house'

This suffix also occurs fossilized in a variant form of the word for 'here': *ayndeb* from *aynde* 'here' + *-mb* OBL after denasalization, also sometimes devoiced finally as *ayndep*.

6.2 Nonverbal clauses

Dixon (2010) divides nonverbal clauses into four types, a classification that I will adopt here. All four of these types are formally distinguished in Kopar. Kopar in contrast to many other languages like English lacks a copula, so these are truly nonverbal clauses in every sense.

6.2.1 Nonverbal clauses of identification

These are clauses in which a subject is identified either as a unique individual, *that guy is my brother*, or as a member of a class of things *Robin is a man*. Some languages distinguish these two uses formally; Kopar does not appear to be one of them (but see section 6.2.3):

(6.41) (a) *nor   yo   kakan   mana*
                nor   yo   kakan   ma-na
                man   DEF older.brother 1SG-POSS
                'the guy is my older brother' (male speaker)

        (b) *mu   nəmandək*
             mu   nəmandək
             3SG woman
             'she is a woman'

6.2.2 Nonverbal clauses of location

These clauses express the location of the subject. The general form of these in Kopar is quite simple, the juxtaposition of the subject noun phrase and a locational postpositional phrase:

(6.42) (a) *taмənd nani sur*
tamənd nani sur
fish pot inside
'the fish is inside the pot'

(b) *nəmbren awr sur*
nəmbren awr sur
pig fire inside
'the pork is in the fire'

However, if the subject is human, either a specific stance verb is used to describe the orientation of the subject's body (6.43a, b) or the general verb of location *ra-* 'be at, stay' is employed:

(6.43) (a) *nor inda sur undasanaraŋgoya*
nor inda sur u-ndasa-n-ar-aŋ-oya
man house inside 3SG-sit-LK-PROG-PRES-3SG
'the man is sitting inside the house'

(b) *mu kay sur urəmara*
mu kay sur u-rəmə-ar-aŋ
3SG canoe inside 3SG-stand-PROG-PRES
'he is standing in the canoe'

(c) *aynde ŋgiraraŋgiya*
aynde ŋgi-ra-ar-aŋ-iya
here 3PC-stay-PROG-PRES-PC
'they (PC) are (staying) here'

(d) *aynde ŋgirarorok*
aynde ŋgi-ra-ar-oro-k
here 3PL-stay-PROG-EXT-FR.PAST
'they (PL) stayed here for a while'

6.2.3 Nonverbal clauses of attribution

Attributive clauses are those in which a property is ascribed to the subject. In Kopar properties are denoted by the adjective or adjectival verbs, and these are the typical heads of nonverbal clauses of attribution in the language. One formal feature that distinguishes this nonverbal clause type from the previous two is that the adjective or adjectival verbs occur with suffixed bound pronominals (optional with third singular):

(6.44) (a) *taməndˌ yo kapu suman(oya)*
tamənd yo kapu suman-(oya)
fish DEF big very-(3SG)
'the fish is very big'

(b) *nəmbren yo ɲaɲjirik(oya)*
nəmbren yo ɲaɲjirik-(oya)
pig DEF small-(3SG)
'the pig is small'

(c) *ma kuŋgoparikənaya*
ma kuŋgoparik-naya
1SG long-1SG
'I am tall'

The border between nonverbal clauses of identification for class membership and those of attribution is a bit hazy in Kopar; consider the following example:

(6.45) *ma nor kuŋgoparikənaya*
ma nor kuŋgoparik-naya
1SG man long-1SG
'I am a tall man'

Example (6.45) could be regard as clause of identification of class membership, 'I am a member of the class of tall men' or attribution 'I am a man who is tall'. The formal structure of an adjectival verb with agreement indicates the second reading is intended, i. e. 'I am a man, I am tall', but the presence of the overt noun without agreement suggests the former. This is a complex predication: *nor kuŋgoparik-naya* man long-1SG expresses both identification and attribution. The property *kuŋgoparik* 'long' is asserted to hold of a subclass of the objects or persons denoted by *nor* 'man', a subclass to which the subject *ma* 1SG belongs.

6.2.4 Nonverbal clauses of possession

Nonverbal clauses of possession are one of the more unusual and typologically interesting features of Kopar grammar. Nonverbal clauses of possession are those in which an object is asserted to be possessed by the subject. Kopar has two constructions which perform this function. The first type simply takes the possessor marked by *-na-k* POSS-NE (see section 4.3.1) and formally verbalizes it via a bound pronominal suffix for the possessor from the ergative set for realized events of Table 6, literally 'subject has X of subject':

(6.46) (a) *puruŋ   manakoya*
           puruŋ   ma-na-k-oya
           betelnut   1SG-POSS-NE-1SG
           'the betelnut is mine' (literally 'I have betelnut of mine')

   (b) *indan   yo   paŋgənakokə*
       indan   yo   paŋgə-na-k-okə
       house   DEF   1PC-POSS-NE-1PC
       'the house is ours (PC)'

Example (6.47) is an interesting variant of this structure in which the possessed object belonging to one person is actually in the possession of another:

(6.47) *manakonduku   moran*
       ma-na-k-onduku   moran
       1SG-POSS-NE-2PC   thing
       'you (PC) have something that is mine' (literally 'you have the thing of mine')

There is a second type of possession clause, and there are two constructions of this type. In the first case the possessed is verbalized directly. A suffix to mark the possessor, again from the set of Table 6, is added to the noun or noun phrase denoting the possessed. If the possessed is expressed by a noun phrase, this suffix can be added to it directly:

(6.48) (a) *indan   kapu   tambək-oya*
           indan   kapu   tambəkoya
           house   big   five-1SG
           'I have five big houses'

(b) *məŋə tamənd sanandəkodudu*
    məŋə tamənd sanandək-ondudu
    3PC   fish   four-3PC
    'they (PC) have four fish'

(c) *mina   indanona*
    mi-na   indan-ona
    2SG-POSS   house-2SG
    'you (SG) have your (SG) house'

(d) *mora munəm   aratəkona*
    mora munəm   aratək-ona
    thing smell   good-2SG
    'you (SG) have a good smell'

(e) *karəkarək manak   nuŋgo aratəkoya*
    karəkarək ma-na-k   nuŋgo aratək-oya
    pillow   1SG-POSS-NE   very   good-1SG
    'I have a very good pillow'

However, if the possessed is represented by a bare noun and optionally if a noun phrase, a suffix *-aŋ* can be added to it. But in this case the bound pronominal suffixes for the possessor are drawn from the nominative set for realized tenses of Table 4. The most likely source for this *-aŋ* is the dative postposition *ŋga*, although I will gloss *-aŋ* as POSS here, given the idiosyncratic form and usage. Support for this claim comes from the Wongan dialect, which in addition to *-aŋ* uses a form *-andək* in this construction, almost certainly from *ndək* PURP, which can often be used interchangeably with *ŋga* DAT in Kopar (see section 6.1.3.2):

(6.49) (a) *məŋə   tamandaŋgiya*
           məŋə   tamənd-aŋ-iya
           3PC   fish-POSS-PC
           'they (PC) have fish'

       (b) *nəmaŋgaya*
           nəma-aŋ-aya
           woman-POSS-1SG
           'I have a wife'

       (c) *məŋə tamənd sanandəkaŋgiya*   (compare (6.48b))
           məŋə tamənd sanandək-aŋ-iya
           3PC   fish   four-POSS-PC
           'they (PC) have four fish'

(d) mi   puruŋ   kombariaŋənaya
    mi   puruŋ   kombari-aŋ-naya
    2SG   betelnut two-POSS-2SG
    'you (SG) have two betelnut'

(e) rarindək   puruŋaŋoya
    rari-ndək   puruŋ-aŋ-oya
    1.day.removed-NR.PAST   betelnut-POSS-3SG
    'he had betelnut yesterday'

With quite complex noun phrases in the absence of *-aŋ*, speakers accept bound pronominal agreement suffixes from either Table 6 for the possessor or Table 4 for the number of the possessed, though this could now be due to confusion or paradigm leveling in the current moribund state of the language:

(6.50) ma   maɲjikap keremən nuŋgo patəndəkoya/iya
       ma   maɲjikap keremən nuŋgo patəndək-oya/iya
       1SG net.bag   three   very   heavy-1SG (Table 6)/PC (Table 4)
       'I have three very heavy netbags'

And there is a not very clear example in one of the narrated texts in which the pronominal is apparently from the dative series; in this case the possessor is a body part, so this could be the result of possessor raising (section 5.5.1):

(6.51) ikunun   munəmbaŋanana
       ikunun   munəmb-aŋ-anana
       sore   smell-POSS-2SG.DAT
       'you (SG) have a smelly sore'

There is yet another type of clause for possession in Kopar in which the possessor is the subject, yet this one may probably be better analyzed as a verbal clause involving a verb equivalent to *have* in English or *punya* 'have' in Indonesian. If so, it is a slightly irregular verb in that it is tenseless, so could be analyzed as a particle, given that tense and mood have been used as diagnostics of verbs in Kopar. On the other hand, it does have one criterial property of verbs: bound pronominals from Table 3 specifiying its subject, in this case, the possessor; Table 13 presents its paradigm (minus the initial /o/ of Table 3 forms, as the possession verb ends in a vowel):

Table 13: Inflection of the Verb of Possession.

	SG	DL	PC	PL
1	yuwa-ya	yuwa-ke	yuwa-k(ə)	yuwa-nde
2	yumu-na	yumu-ko	yumu-nduku	yumu-ndo
3	yumu-nda	yumu-ndə	yumu-dudu	yumu-ndu

Here are a couple of examples of its use:

(6.52) (a) ke   puruŋ   yuwake
　　　　　ke   puruŋ   yuwa-oke
　　　　　1DL betelnut have-1DL
　　　　　'we (DL) have betelnut'

　　　(b) ruaŋ   yumukwe
　　　　　ruaŋ   yumu-oko-e
　　　　　coconut have-2DL-Q
　　　　　'do you (DL) have a coconut?'

Lack of possession is expressed in the expected way, by the use of *kay-* NEG. As expected, the bound pronominals from Table 3 for unrealized tenses and moods occur, even in the presence of *-aŋg*, but again 'climb' to attach to the negator:

(6.53) mu   puruŋang   kaynda
　　　　mu   puruŋ-ang   kay-onda
　　　　3SG betelnut-POSS NEG-3SG
　　　　'he has no betelnut'

Chapter 7
Interclausal relations

Kopar makes use of both subordination and coordination to link clauses together. Subordinated clauses come in two types, finite and non-finite. Non-finite subordinate clauses are divided into two kinds: infinitives, of which there are again two types, those marked by the dative postposition *ŋga* and those with the purposive postposition *ndək*, and nominalizations. Finite subordinate clauses correspond to adverbial subordinate clauses of more familiar languages like English, French or Tok Pisin. Coodinated clauses also come in two kinds: juxtaposed clauses with full inflected verbs in both clauses and the typical Papuan clause chaining constructions in which dependent clauses with morphogically stripped down verbs are joined to an independent clause headed by a fully inflected verb.

7.1 Non-finite constructions

There are two types of infinitives in Kopar, and both are used for subordinate clauses which express events that are not yet realized. They can express the purpose for which another event is carried out, and this is especially common with motion verbs in the main clause, or those which denote events which are desired or shunned.

7.1.1 Dative infinitive constructions

These infinitive constructions always occur with the dative postposition *ŋga*, while the verb is usually unmarked or very occasionally marked with a suffix *-n*, which may indicate nominalization, as the presence or absence of a final /n/ is diagnostic of nouns (section 4.1). Dative infinitives never have tense or person-number marking for core arguments. With motion verbs in the main clause, dative infinitives express the event which is the intended goal of the motion:

(7.1) (a) mu   obida   pipikambu   ŋga
         mu   o-mbi-onda   pipikambu   ŋga
         3SG   go-IM.FUT-3SG   fish.with.hook   DAT
         'she should go fishing'

(b) | rari | onduk | kaynda | pipikambu | ŋga |
    | rari | o-onduk | kay-onda | pipikambu | ŋga |
    | 1.day.removed | go-FUT | NEG-3SG | fish.with.hook | DAT |
    'she won't go fishing tomorrow'

But they occur with other types of main verbs as well to express an intended event as the goal of the action of the main verb:

(7.2) (a) | mana | ŋga | awr | kuteka |
          | ma-na | ŋga | awr | ku-t-e-ka |
          | 1SG-POSS | DAT | fire | TR.IMP-CAUS-come-2SG |
          | mumora | si | ŋga |
          | mə-mora | si | ŋga |
          | eat-thing | make | DAT |
          'bring me fire to cook food!'

   (b) | urəmanumbwak | kar | erakot | ŋga |
       | u-rəmə-anumbwa-k | kar | e-ra-kot | ŋga |
       | 3SG-stand-3PL.DAT-FR.PAST | feast | come-stay-carry | DAT |
       'he stood up for them (PL) in order to bring the feast'

The most common use of dative infinitives is to express desirable or undesirable, but as yet unrealized, events. Thus, they heavily feature in desiderative constructions or their opposites. There are two ways to express desiderative constructions with dative infinitives. One is to combine them with the main verb *si-* 'do, make, become, happen, feel':

(7.3) (a) | mu | pipikambu | ŋga | o | ŋga | usindukoya |
          | mu | pipikambu | ŋga | o | ŋga | u-si-ndək-oya |
          | 3SG | fish.with.hook | DAT | go | DAT | 3SG-feel-NR.PAST-3SG |
          'she wanted to go fishing'

   (b) | o | ŋga | ŋgisikəŋgaya |
       | o | ŋga | ŋgi-si-k-ŋgaya |
       | go | DAT | 3PL-feel-FR.PAST-3PL |
       'they (PL) wanted to go'

   (c) | ambisen | ukandəksemakoya |
       | ambisen | u-kandək-se-ma-k-oya |
       | daughter | 3SG-sleep-NIGHT-DUR-FR.PAST-3SG |

miɲjir   pra   ŋga   usianakoya
miɲjir   pra   ŋga   u-si-ana-k-oya
urine   excrete   DAT   3SG-do-3SG.DAT-FR.PAST-3SG
'the daughter was sleeping during the night and wanted to urinate'

(d) nana   e   ŋga   usianakoya
    nana   e   ŋga   u-si-ana-k-oya
    mama   come   DAT   3SG-do-3SG.DAT-FR.PAST-3SG
    'mother wanted to come'

This construction is only used in in the realized tenses. Otherwise the verb *ri-* 'want' is used with bound pronominal affixes for the unrealized tenses and moods, although the verb can be omitted with the same meaning (7.4e):

(7.4) (a) kayn   paka   ŋga   rine
          kayn   paka   ŋga   ri-ona-e
          canoe   carve   DAT   want-2SG-Q
          'do you (SG) want to carve a canoe?'

     (b) nəmbren   nambrat   ŋga   rində
         nəmbren   nambrat   ŋga   ri-ondə
         pig   spear   DAT   want-3DL
         'they (DL) want to spear a pig'

     (c) arəm   kiri   ŋga   riya
         arəm   ki-ri   ŋga   ri-oya
         water   bathe-DOWN   DAT   want-1SG
         'I want to bathe'

     (d) Keli   nda   sara   kata   ŋga   riya
         Keli   nda   sara   kata   ŋga   ri-oya
         Kelly   COM   speak   say   DAT   want-1SG
         'I want to speak with Kelly'

     (e) ma   pipikambu   ŋga
         ma   pipikambu   ŋga
         1SG   fish.with.hook   DAT
         'I want to fish'

This desiderative use of *ri-* 'want' plus dative is not restricted to infinitives; it is also possible with nouns:

(7.5) (a) *mi sokayn ŋga rine*
　　　　mi　sokayn　ŋga　ri-ona-e
　　　　2SG　tobacco　DAT　want-2SG-Q
　　　　'you (SG) want cigarettes?'

　　(b) *puruŋ ŋga ri-kə*
　　　　puruŋ　ŋga　ri-okə
　　　　betelnut　DAT　want-1PC
　　　　'we (PC) want betelnut'

The verb *nda-* 'don't want' is used in the same way to indicate events that are undesirable (7.6a), but it is also possible in these constructions to omit the dative postposition (7.6b):

(7.6) (a) *arəm kiri ŋga ndaya*
　　　　arəm　ki-ri　　　　ŋga　nda-oya
　　　　water　bathe-DOWN　DAT　don't.want-1SG
　　　　'I don't want to bathe'

　　(b) *ma mə ndaya*
　　　　ma　mə　nda-oya
　　　　1SG　eat　don't.want-1SG
　　　　'I don't want to eat'

It is possible that the use of the dative postposition here is due to Tok Pisin influence *mi les long waswas* 'I don't want to bathe' and that the usage without it is the older one. On the other hand, it never seems omissible in the positive desideratives of (7.3)-(7.4).

Desideratives can also be expressed as a complement of the noun *nunon* 'mind, desire, feelings', but again with a dative infinitive:

(7.7) *pipikambu ŋga nunon kapuya*
　　　pipikambu　　ŋga　nunon　kapu-oya
　　　fish.with.hook　DAT　desire　big-1SG
　　　'I have a big desire to fish'

Finally, dative infinitives are used with other types of unrealized intended events, such as with modals:

(7.8) (a) taynonda   indan undi   ŋga
          tayn-onda   indan undi   ŋga
          ABIL-3SG   house build DAT
          'he can build a house'

   (b) sina   mumora   nambri si   ŋga tayn   kayndu
       sina   mə-mora   nambri si   ŋga tayn   kay-ondu
       daytime   eat-thing   eye   do   DAT   ABIL NEG-3PL
       'they (PL) can't find food in the daytime'

The sequence *ŋga* DAT plus *ri-* 'want' has been re-analyzed into an alternative dative postposition to *ŋga* DAT that can be used both with noun or pronominal complements (7.9a) and for dative infinitives (7.9b, e):

(7.9) (a) nor   puruŋ   mana   ŋari   mbu-t-e-n-aŋg-oya
          nor   puruŋ   ma-na   ŋari   mbu-t-e-n-aŋg-oya
          man   betelnut 1SG-POSS DAT   3.ERG-CAUS-come-LK-PRES-3SG
          'the man gives me a betelnut'

   (b) mu   tamənd   mbuketamaŋgoya   mə   ŋari
       mu   tamənd   mbu-ketam-aŋg-oya   mə   ŋari
       3SG   fish   3.ERG-bring.ashore-PRES-3SG   eat   DAT
       'she caught fish to eat'

   (c) kutururujabiduku   paret   ŋari
       ku-t-ruruɲja-mbi-onduku   paret   ŋari
       TR.IMP-CAUS-shake-IM.FUT-2PC   outside DAT
       'you (PC) shake (it) until (it comes) out'

   (d) ŋerasakəŋgaya   atik   ŋari
       ŋgi-e-ra-sa-k-ŋgaya   atik   ŋari
       3PL-come-stay-IN-FR.PAST-3PL   smoke   DAT
       'they (PL) came and stayed inside the smoke'

   (e) ma   ruaŋ   mə   ŋari
       ma   ruaŋ   mə   ŋari
       1SG   coconut eat DAT
       'I want to eat a coconut'

7.1.2 Purposive infinitive constructions

These infinitive constructions are marked with the purposive postposition *ndək*. It is often interchangeable with *ŋga* or *ŋgari* (compare (7.10a) with (7.9b)):

(7.10) (a) mu   tamənd   mbuketamaŋgoya   mə   ndək
            mu   tamənd   mbu-ketam-aŋg-oya   mə   ndək
            3SG fish   3.ERG-bring.ashore-PRES-3SG eat PURP
            'she caught fish for eating'

(b)   Wak   ŋga   tra   ndək   pək
      Wak   ŋga   tra   ndək   pək
      PN   DAT   dance PURP platform
      'a platform for dancing for Wak'

In purposive infinitives *ndək* PURP sometimes co-occurs with an immediately preceding prosodic particle *o*, which has become bound and causes vowel harmony of the /ə/ of the postposition. The resulting *onduk* also can become bound to its complement and undergo denasalization if the conditions are met (7.11a). This accretion results in homophony between the postposition in purposive infinitives and the future tense suffix. This allomorph of the purposive postposition is also interchangeable with *ŋga* DAT (7.11b):

(7.11) (a) mu   naŋgep   mbwerakotundukoya
            mu   naŋgep   mbu-e-ra-kot-ndək-oya
            3SG house.post 3.ERG-come-stay-carry-NR.PAST-3SG
            inda   undioduk
            Inda   undi-ondək
            house build-PURP
            'he brought a post to build a house'

(b)   mu   naŋgep   mbwerakotundukoya
      mu   naŋgep   mbu-e-ra-kot-ndək-oya
      3SG house.post 3.ERG-come-stay-carry-NR.PAST-3SG
      inda   undi   ŋga
      inda   undi   ŋga
      house build DAT
      'he brought a post to build a house'

7.1.3 Nominalization constructions

Only a few examples of nominalizations were identified in the corpus. Nominalizations are marked simply by adding the noun marking suffix *-n* to a verb root. There was only one example of an overt subject with these nominalizations, and, as in many languages, it undergoes genitivization (7.12e). But objects in nominalizations remain bare, not undergoing genitivization, unlike English:

(7.12) (a) sokay   mən   mora   sin   makona
           sokay   mə-n   mora   si-n   mak-ona
           tobacco   eat-NMLZ thing do-NMLZ   bad-2SG
           'smoking is bad for you' (literally 'you have a bad deed, smoking tobacco')

   (b) mu   nde   tran   yo   kwarik   ena
       mu   nde   tra-n   yo   kwarik   ena
       3SG   like.this   dance-NMLZ   DEF   side   PROX.SG
       'he on this side, the dance like this'

   (c) [Sapend ndasiarək]   mora   sin   yo
       [Sapend nda-si-ar-k]   mora   si-n   yo
       PN   NOW-do-PROG-FR.PAST   thing do-NMLZ   DEF
       mbusamaytarək
       mbu-samayt-ar-k
       3.ERG-see-PROG-FR.PAST
       'they were watching the action that Sapend was doing'

   (d) saran   gidukoya
       sara-n   ŋgi-ndə-k-oya
       report-NMLZ   3PL-hear-FR.PAST-3
       'they (PL) listened to what he said'

   (e) saran   ombe   kaynda   kata   muna
       sara-n   ombe   kay-onda   kata   mu-na
       report-NMLZ   INDEF   NEG-3SG   speak   3SG-POSS
       'he did not say any report of his' (Tok Pisin *em i no tokim wanpela tok bilongen*)

   (f) apasen   minak   o   mora   sin   ombe
       apasen   mi-na-k   o   mora   si-n   ombe
       grandmother   2SG-POSS-NE   thing   happen-NMLZ   INDEF
       '(to) your (SG) grandmother, some event (happened)'

7.2 Finite constructions

7.2.1 Finite subordinate clauses

Finite subordinate clauses in Kopar appear to be limited to expressing temporal, hypothetical, possible or counterfactual conditions on a possible resulting event. Semantic relations expressed by temporal adverbial clauses of more familiar languages like sequence are generally expressed in Kopar by clause coordination (section 7.2.2) or clause chaining (section 7.2.3). The protasis or finite subordinate clause expresses the condition, and the apodosis or main clause describes the result that could or should have occurred or failed to occur in the case of counterfactuals. Events in finite subordinate clauses are expressed with the irrealis suffix -*k* (homophonous with the far past) plus the oblique suffix -*mb* often followed by the prosodic particle *o*. Verbs in the protasis of the subordinated hypothetical or conditional clauses do not occur with bound suffixed pronominals, while those of the apodosis always occur with bound pronominals from the series for unrealized tenses or moods:

(7.13) (a) ŋgu   mokəmb   o   mana   awr sur o
ŋgu   mə-o-k-mb   o   ma-na   awr sur o
2PC   eat-go-IRR-OBL   1SG-POSS fire in
'when you (PC) eat and go, (leave) mine in the fire!'

(b) ma   arəm yo məkəmb   naŋgun bijabin   nda
ma   arəm yo mə-k-mb   naŋgun bijabi-n   nda
1SG water DEF eat-IRR-OBL skin light-NMLZ COM
'if I drink the water, (my) skin (will be) light (not heavy)'

(c) okəmb   o   samerarəkəmbaya
o-k-mb   o   samer-arək-mbaya
go-IRR-OBL   wait-PROHIB-DL
'when going, you (DL) can't wait'

(d) ke   okəmb   o nəmbən
ke   o-k-mb   o nəmbən
1DL go-IRR-OBL garamut.drum
kuturikermake
ku-t-riker-ma-oke
TR.IMP-CAUS-get.up-AFTERNOON-1DL
'when we (DL) go, let us (DL) sound the garamut in the afternoon'

(e) *mana   ŋga   ekəmb   o   ma   bagariondukoya*
    ma-na   ŋga   e-k-mb   o   ma   mbaŋgari-onduk-oya
    1SG-POSS   DAT   come-IRR-OBL   1SG kill-FUT-1SG
    'if he comes for me, I will kill him'

(f) *sokay   məkəmb   o   mak   simbina*   (compare
    sokay   mə-k-mb   o   mak   si-mbi-ona   7.11a)
    tobacco   eat-IRR-OBL   bad   become-IM.FUT-2SG
    'if (you) smoke, you (SG) will become bad'

(g) *ma   awr   təkarikəmb   o   awr   repesandə   kay-nda*
    ma   awr   t-kari-k-mb   o   awr   repesa-andə   kay-nda
    1SG   fire   CAUS-put-IRR-OBL   fire   ignite-PFV   NEG-3SG
    'if I start a fire, it won't ignite'

(h) *ma   anumb   sumbekəmb   o   naɲjen   mana   anumb*
    ma   anumb   sumbe-k-mb   o   naɲjen   ma-na   anumb
    1SG   sago.palm   plant-IRR-OBL   child   1SG-POSS   sago.palm
    mondukududu
    m-onduk-undundu
    eat-FUT-3PC
    'if I plant a sago palm, my children will eat sago'

Counterfactuals have much the same structure, but instead of the oblique suffix *-mb*, they take a postposition *bisi*, and the apodosis clause is in a past tense. Compare the examples in (7.14) for the difference between hypothetical/conditional subordinate clauses (7.14a) and counterfactuals (7.14b):

(7.14) (a) *mi   mu   bagarikəmb   o   rari   mi   sara-n*
           mi   mu   mbaŋgari-k-mb   o   rari   mi   sara-n
           2SG   3SG   kill-IRR-OBL   tomorrow   2SG   report-NMLZ
           erakotondukona
           e-ra-kot-onduk-ona
           come-stay-carry-FUT-2SG
           'if you (SG) kill him, tomorrow you (SG) will bring a report (i. e. gossip)'

       (b) *mi   rarindək   mu   bagari   bisi*
           mi   rari-ndək   mu   bagari   bisi
           2SG   1.day.removed-NR.PAST   3SG   kill   COUNTERFACT

> mi saran məŋgəna indaerakotndəkəna
> mi sara-n məŋgə-na i-nda-e-ra-kot-ndək-na
> 2SG report-NMLZ 3PC-POSS 2-NOW-come-stay-carry-NR.PAST-2SG
> 'if you (SG) had killed him yesterday, you (SG) would have brought their (PC) talk by now'

Note that the apodosis clause of counterfactual sentences has a past tense specification, a tense of the realized set; the verbs of these clauses occur with the bound nominative pronominals from Table 4, albeit the suffixal part is truncated in form by the loss of the final syllable -ya. This is somewhat unexpected because the semantics of these apodosis clauses is to describe events which could have occurred, but did not in fact occur, in other words, unrealized events, yet the formal realization by suffixes from a realized tense seems to override that. Here are two other examples of counterfactual subordinate clauses:

(7.15) (a) mu indan yo aratək undi bisi
  mu indan yo aratək undi bisi
  3SG house DEF good build COUNTERFACT
  akən rotəndək kaynda
  akən rot-ndək kay-onda
  rain fall-NR.PAST NEG-3SG
  'if he had built the house well, rain wouldn't have fallen in'

(b) ma patemba skul o bisi
  ma pate-mba skul o bisi
  1SG before-LOC school go COUNTERFACT
  titser mandakusaməndəkəna
  titser ma-nda-kusa-am-ndək-na
  teacher 1SG-NOW-go.out-DETR-NR.PAST-1SG
  'if I had gone to school before, I would have graduated as a teacher by now'

Finite subordinate clauses can also be marked with the adverbial suffix -ndi, and again this suffix co-occurs with -k IRR. Clauses with -ndi ADV indicate a simultaneous temporal relationship between the two clauses and are best translated as 'while'. It is not entirely clear whether these constructions should be analyzed as finite adverbial subordinate clauses or clause chaining constructions (section 7.2.3), but given that, like the other more clearly subordinate clauses discussed in this section, clauses marked by -ndi always co-occur with -k IRR, it seemed prudent to discuss them here. On the other hand, the fact that they display 'argument bleeding', in which the argument of a verb is realized in a clause with a verb that does not semantically govern it, a behavior they share with clause chaining constructions (example 7.27), may suggest that a clause chaining analysis may be correct for them. Examples follow:

(7.16) (a) mu   karokəndi   nəmbre ŋga nambrin
mu   kar-o-k-ndi   nəmbre ŋga nambrin
3SG walk-go-IRR-ADV pig   DAT eye
utukaməndukoya
u-t-kam-ndək-oya
3SG-CAUS-arrive-NR.PAST-3SG
'walking about, he kept his eye on the pig'

(b) mu kay sur rəməkəndi uren
mu kay sur rəmə-k-ndi uren
3SG canoe inside stand-IRR-DEP crocodile
mbunambrataŋgoya
mbu-nambrat-aŋ-oya
3.ERG-spear-PRES-3SG
'standing up in the canoe, he spears the crocodile'

(c) mbutururuɲjarəkududu   paret   ŋga
mbu-t-ruruɲja-ar-k-ududu   paret   ŋga
3.ERG-CAUS-shake-PROG-IRR-3PC outside DAT
sikəndi
si-k-ndi
make-FR.PAST-ADV
'they (PC) were shaking it, making it (come) out'

(d) aynde sara katakatakəndi   mbukinimbəkəto
aynde sara kata+kata-k-ndi   mbu-kinimb-k-to
here   report speak.RED-IRR-ADV 3.ERG-fasten-FR.PAST-DEP
mbusambotəkəto   urarək
mbu-sambo-t-k-to   u-ra-ar-k
3.ERG-leave-APPL-FR.PAST-DEP 3SG-stay-PROG-FR.PAST
'talking like that here, he tied him up and left him and he remained'

7.2.2 Coordination of full independent clauses

Kopar is unusual among Papuan languages, and perhaps more generally among languages with a verb final typology, in that coordination of full independent clauses is at least as common a way of coordinating clauses as the clause chaining structures of section 7.2.3, and maybe even more common (though it does share this property with the languages of the southern lowlands of Papua Province in Indonesia and the Western Province of Papua New Guinea). This also could be due to its overall polysynthetic typology, as it is well known that polysynthetic languages often favor simple coordination or parataxis as a way of linking clauses together (Baker 1996; Mithun 1984). Indeed, coordinate full clauses seem unusually prominent in Kopar when compared with more familiar Papuan languages such as those of the Trans New Guinea family and even its sister language Yimas. In this type of coordination, all verbs are fully inflected and can stand by themselves as independent utterances. Note that coordinated full independent clauses can be used when there are subjects with the same referents across the clauses (7.17b, c, d, e) and well as when different (7.17a, b, d, f):

(7.17) (a) mi   ma-na   nəman   nambratondukona
           mi   ma-na   nəman   nambrat-onduk-ona
           2SG   1SG-POSS   woman   spear-FUT-2SG
           ma   mi-na   nəman   nambratondukoya
           ma   mi-na   nəman   nambrat-onduk-oya
           1SG   2SG-POSS   woman   spear-FUT-1SG
           'you (SG) will spear my wife and I will spear your (SG) wife'

       (b) Wak   aynde   usaranumbwakoya   sara-n
           Wak   aynde   u-sara-anumbwa-k-oya   report-n
           PN   here   3SG-report-3PL.DAT-FR.PAST-3SG   talk-NMLZ
           gi-du-k-oya   naŋgrin   ŋgiprəkək
           ŋgi-ndə-k-oya   naŋgrin   ŋgi-prək-k
           3PL-hear-FR.PAST-3   spear   3PL-make-FR.PAST
           'Wak reported it to them (PL) here and they (PL) listened to his report and they (PL) made spears'

       (c) o   mobido   kumuobido
           o   ma-o-mbi-ondo   ku-mə-o-mbi-ondo
           2PC   ITR.IMP-go-IM.FUT-2PL   TR.IMP-eat-go-IM.FUT-2PL
           porakay   yo
           porakay   yo
           path   DEF
           'you (PL) go and you (PL) eat while going along the path'

(d) kakandək   yo   mbutumuranak
    kakandək   yo   mbu-t-mur-ana-k
    older.same.sex.sibling DEF 3.ERG-COM-afraid-3SG.DAT-FR.PAST
    nda itəmandək   enana   mbukamanak
    nda itəmandək   enana   mbu-kam-ana-k
    and younger.sister♀ PROX.SG 3.ERG-give-3SG.DAT-FR.PAST
    mbutakumbanak   mu-na   ŋga
    mbu-takumb-ana-k   mu-na   ŋga
    3.ERG-split-3SG.DAT-FR.PAST 3SG-POSS DAT
    'the older sister was afraid of that and this younger sister gave it to her and split it for her'

(e) apikor   aratək   mbutotanak
    apikor   aratək   mbu-t-o-ta-ana-k
    necklace   good   3.ERG-CAUS-go-OUT-3SG.DAT-FR.PAST
    mbukamanak
    mbu-kam-ana-k
    3.ERG-give-3SG.DAT-FR.PAST
    'she took out a good necklace for her and gave it to her'

(f) akən abeb   apasen   yowa   onakoya   səmbər
    akən abeb   apasen   yowa   o-na-k-oya   səmbər
    sun   other grandmother DIST   go-UP-FR.PAST-3SG forest
    nda yan   kombari rama   kaka
    nda yan   kombari rama   kaka
    and 3DL   two   younger.brother♂ older.same.sex.sibling
    nda mbə naŋgatək tamənd   mbiruesirik
    nda mbə naŋgatək tamənd   mbi-ru-e-siri-k
    and 3DL current   fish   3DL-shoot-come-go.down-FR.PAST
    'another day, that grandmother went to the forest, and two papas, a younger brother and an older brother, came down on the current spearing fish'

One environment that favors this construction is when there is a major shift of tenses between the two clauses, as in this example:

(7.18) nor   nəmbren   mbunambrataŋgoya   tumbumadududu
       nor   nəmbren   mbu-nambrat-aŋg-oya   t-mbu-mə-andə-ndundu
       man   pig   3.ERG-spear-PRES-3SG   PFV-3.ERG-eat-PFV-3PC
       'the man shoots the pig and they have eaten it'

Sometimes coordinated full clauses express a causative relationship where the second clause is the causal motivation for the action of the first:

(7.19) ma ruaŋ mə ŋgari ruaŋ aratəkoya
      ma ruaŋ mə ŋgari ruaŋ aratək-oya
      1SG coconut eat DAT coconut good-1SG
      'I want to eat the coconut because I have a good coconut'

Coordination of full clauses with independent verbs is also used when the mood of the two clauses is different (7.20a) or their relationship is disjunctive (7.20b):

(7.20) (a) mi pipikambu   ŋga o ndana
          mi pipikambu   ŋga o nda-ona
          2SG fish.with.hook DAT go PROHIB-2SG
          arəm kiri   ŋga moka
          arəm ki-ri   ŋga ma-o-ka
          water bathe-DOWN DAT ITR.IMP-go-2SG
          'don't go fishing, go bathe!'

     (b) mi ruaŋ   sayt ŋga rineee
          mi ruaŋ   sayt ŋga ri-na-eee
          2SG coconut pick DAT want-2SG-Q
          arəm kiri   ŋga
          arəm ki-ri   ŋga
          water bathe-DOWN DAT
          'do you (SG) want to pick a coconut or bathe?

As example (7.20b) demonstrates, disjunctive questions are expressed by questioning the first coordinated clause and lengthening the question marker -*e*.

7.2.3 Clause chaining

Clause chaining is a type of coordination in which clauses are in an unequal relationship. Dependent clauses are typically headed by morphologically stripped down or dependent verbs which co-occur with a clause headed by a fully inflected or independent verb. Dependent clauses and verbs are so named because they are dependent on independent clauses and verbs for their specifications of certain grammatical categories like tense or mood. In comparison to other Papuan languages, dependent verbs in Kopar are not very impoverished, and only minimally distinguished from independent verbs. Dependent verbs still occur with tense marking and also with bound pronominals for core arguments, although normally only with prefixes; the dependent verb marking suffixes usurp the position for the bound pronominal suffixes. Except for the presence of the dependent suffix, dependent verbs in Kopar are indistinguishable from independent verbs, as independent verbs in ongoing text also often lack pronominal agreement suffixes and terminate in the tense suffix (see examples in (7.17). Furthermore, Kopar dependent clauses do not require coreferential subjects like many other Papuan languages; there is no switch reference morphology nor any constraint that the subjects of verbs in clause chaining constructions have the same referent. There are two common markers of dependent verbs in clause chaining constructions in Kopar: *-to* and *-andə*. The more common is the suffix *-to*, which simply marks that the verb is dependent, i. e. cannot stand on its own as a sentence:

(7.21) (a) nəmand yowa surun nda usianak
nəmand yowa suru nda u-si-ana-k
woman DIST belly COM 3SG-do-3SG.DAT-FR.PAST
ŋgimaŋgarorokəto nəmand
ŋgi-ma-ŋga-ar-oro-k-to nəmand
3PC-row-go.into-PROG-EXT-NR.PAST-DEP woman
umbeanakəto sen Wak Wak yowa
u-mbe-ana-k-to sen Wak Wak yowa
3SG-bear-3SG.DAT-FR.PAST-DEP son PN PN DIST
'the woman was pregnant, and they (PC) rowed for a while, and the woman gave birth to a son Wak, that Wak'

(b) ŋgiokəto rakamdan awr
ŋgi-o-k-to rakamdan awr
3PC-go-FR.PAST-DEP night fire
mbukarisekududu
mbu-kari-se-k-ududu
3.ERG-put-NIGHT-FR.PAST-3PC
'they (PC) went and they (PC) set a fire at night'

(c) uramakəto mbu ŋgiok
u-ra-ma-k-to mbu ŋgi-o-k
3SG-stay-DUR-FR.PAST-DEP 3PL 3PL-go-FR.PAST
'he continued to stay and they (PL) went'

Note that in (7.21b) the subjects of the two clauses are coreferential, while in (7.21c) they are disjoint in reference. But the structure of the two sentences is the same;

there is no switch reference system in the clause chaining constructions of Kopar, as many examples of (7.21) illustrate.

(d) *mbumokəto*　　　　　　*mbuna*　*ta*　*ukrarik*
mbu-mə-o-k-to　　　　　　mbu-na　ta　u-krari-k
3.ERG-eat-go-FR.PAST-DEP　3PL-POSS　from　3SG-hide-FR.PAST
'he ate while going and hid from them (PL)'

(e) *mbu*　*ŋgiok*　　　　　*mu*　*ndetana*
mbu　ŋgi-o-k　　　　　mu　ndetana
3PL　3PL-go-FR.PAST　3SG　alone
urarorokəto　　　　　　　　　　*nəmandək*　*kombar*
u-ra-ar-oro-k-to　　　　　　　　nəmandək　kombar
3SG-stay-PROG-EXT-FR.PAST-DEP　woman　　　two
mbutəmek
mbu-təme-k
3.ERG-tell-FR.PAST
'they (PL) went and he alone stayed for a while and he told two women'

(f) *nimbep*　*napar*　*sur*　*o*　*ndək*　*urotanakəto*
nimbep　napar　sur　o　ndək　u-rot-ana-k-to
spear　　hand　　in　　　　PURP　3SG-fall-3SG.DAT-FR.PAST-DEP
nda　*urarianak*
nda　u-rari-ana-k
and　3SG-cry-3SG.DAT-FR.PAST
'the spear dropped from his hand and he cried over her'

(g) *yan*　*enana*　　*mbukamendakəto*
yan　　enana　　mbu-kame-anda-k-to
papa　PROX.SG　3.ERG-call.out-3DL.DAT-FR.PAST-DEP
mbisirikəto　　　　　　　　*mbuturarindak*
mbi-siri-k-to　　　　　　　mbu-t-rari-anda-k
3DL-go.down-FR.PAST-DEP　3.ERG-COM-cry-3DL.DAT-FR.PAST
'this papa called out to them (DL) and they (DL) came down and he cried over them (DL)'

(h) *indaimbot*　*mbuturəməkəto*
inda-imbot　mbu-t-rəmə-k-to
house-nose　3.ERG-CAUS-stand-FR.PAST-DEP

mbukinimbəkəto   mbusambotək
mbu-kinimb-k-to   mbu-sambo-t-k
3.ERG-fasten-FR.PAST-DEP 3.ERG-leave-APPL-FR.PAST
'he stood her up and fastened her to the gable and left her'

(i) rakamda   nana   ndurum   yowa   onakəto
    rakamda   nana   ndurum   yowa   o-ana-k-to
    night   mama   ghost   DIST   go-3SG.DAT-FR.PAST-DEP
    ambisen   mbutusisek
    ambisen   mbu-t-si-se-k
    daughter 3.ERG-CAUS-feel-NIGHT-FR.PAST
    'at night that mother's ghost went to her and stirred the daughter'

(j) ende   sara   katakatakəndi   mbukinimbəkəto
    ende   sara   kata+kata-k-ndi   mbu-kinimb-k-to
    so   report   speak.RED-IRR-ADV   3.ERG-fasten-FR.PAST-DEP
    mbusambotəkəto   urarək
    mbu-sambo-t-k-to   u-ra-ar-k
    3.ERG-leave-APPL-FR.PAST-DEP 3SG-stay-PROG-FR.PAST
    'while talking like that, he tied him up and left him and he remained (there)'

(k) endeŋ   ŋgisiarorokəto
    endeŋ   ŋgi-si-ar-oro-k-to
    like.this 3PL-do-PROG-EXT-FR.PAST-DEP
    nda   mbutundatəkondu
    nda   mbu-t-ndat-k-ondu
    and   3PL-CAUS-know-FR.PAST-3PL
    'they (PL) did like this and they (PL) understood'

Dependent verb marked with -*to* are the typical way tail-head linkages (de Vries 2005) are formed in Kopar. The final independent verb of the previous sentence is repeated as a dependent verb marked with -*to* DEP in the first clause of the next sentence, though in comparison to Trans New Guinea languages, especially those of the Highlands, Kopar sentences are short, not more than three or four clauses:

(7.22) (a) akən   usisenumbwak.
           akən   u-si-se-anumbwa-k
           sun   3SG-do-NIGHT-3PL.DAT-FR.PAST

```
        akən      usisenumbwakəto
        akən      u-si-se-anumbwa-k-to
        sun       3SG-do-NIGHT-3PL.DAT-FR.PAST-DEP
        ŋgitrasek.                    ŋgitrasekəto
        ŋgi-tra-se-k.                 ŋgi-tra-se-k-to
        3PL-dance-NIGHT-FR.PAST       3PL-dance-NIGHT-FR.PAST-DEP
        Jina    kandəknambrin    usianak
        Jina    kandək-nambrin   u-si-ana-k
        PN      sleep-eye        3SG-do-3SG.DAT-FR.PAST
```
'night fell on them (PL). Night fell on them (PL) and they danced. They danced and Jina felt sleepy'

(b) ```
    urikeranəmbakoya   mbikosirik.
    u-riker-anəmba-k-oya   mbi-kosiri-k.
    3SG-get.up-3DL.DAT-FR.PAST-3SG   3DL-go.across-FR.PAST
    mbikosirikəto   Kotokari   arata
    mbi-kosiri-k-to   Kotokari   arata
    3DL-go.across-FR.PAST-DEP   PN   good
    urəmanak
    u-rəmə-ana-k
    3SG-stand-3SG.DAT-FR.PAST
    ```
'he came up toward them (DL) and they (DL) came across. They (DL) came across and Kotokari stood up (facing) him well'

(c) ```
    Kandok    aynde    ŋgisirik.                aynde
    Kandok    aynde    ŋgi-siri-k.              aynde
    PN        here     3PL-go.down-FR.PAST      here
    ŋgisirikəto                    mbuna     nəmbən   kapu
    ŋgi-siri-k-to                  mbu-na    nəmbən   kapu
    3PL-go.down-FR.PAST-DEP        3PL-POSS  garamut  big
    mbutok,                   Ŋikman
    mbu-t-o-k,                Ŋikman
    3.ERG-CAUS-go-FR.PAST     PN
    ```
'they came down to Kandok here. They (PL) came down here and took their (PL) big garamut, Ŋikman (name of garamut drum)'

(d) ```
    usaranumbwakoya   saran
    u-sara-anumbwa-k-oya   sara-n
    3SG-report-3PL.DAT-FR.PAST-3SG   report-NMLZ
    ```
    ```

	gi-du-k-oya	naŋgrin	ŋgiprəkək.	naŋgrin
	ŋgi-ndə-k-oya	naŋgrin	ŋgi-prək-k.	naŋgrin
	3PL-hear-FR.PAST-3SG	spear	3PL-make-FR.PAST	spear
	ŋgiprəkəkəto	mayn	yo	runumbwak
	ŋgi-prək-k-to	mayn	yo	ru-anumbwa-k
	3PL-make-FR.PAST-DEP	male	DEF	shoot-3PL.DAT-FR.PAST

'he reported to them (PL) and they (PL) listened to his report and fashioned spears. They (PL) fashioned spears and the man shot them (PL)'

(e) mu   mbiok.   okəto
    mu   mbi-o-k.   o-k-to
    3SG AGAIN-go-FR.PAST   go-FR.PAST-DEP
    naŋgun   mak yo   mbumetik
    naŋgun   mak yo   mbu-meti-k
    skin   bad DEF 3.ERG-shed-FR.PAST

'he went again. He went and he removed the bad skin'

(f) mbə naŋgatək tamənd mbiruesirik.
    mbə naŋgatək tamənd mbi-ru-e-siri-k.
    3DL current   fish   3DL-shoot-come-go.down-FR.PAST
    mbiesirikəto   nəma   sur
    mbi-e-siri-k-to   nəma   sur
    3DL-come-go.down-FR.PAST-DEP   midstream in
    aynde sen yo   nambrin onakoya
    aynde sen yo   nambrin o-ana-k-oya
    here   son DEF   eye   go-3SG.DAT-FR.PAST-3SG
    mbusamaytandakoya
    mbu-samayt-anda-k-oya
    3.ERG-see-3DL.DAT-FR.PAST-3SG

'they (DL) were spearing fish on the current as they (DL) came downsteam. They (DL) came downstream and the son looked here in the middle of the river and saw them (DL)'

(g) nana   nda   aynde   ŋgirarorok.
    nana   nda   aynde   ŋgi-ra-ar-oro-k.
    mama   COM   here   3PL-stay-PROG-EXT-FR.PAST
    ŋgirarorokəto   sen kapu
    ŋgi-ra-ar-oro-k-to   sen kapu
    3PL-stay-PROG-EXT-FR.PAST-DEP   son big

>   *usianak*
>   u-si-ana-k
>   3SG-become-3SG.DAT-FR.PAST
>   'they remained here with mama for a while. They remained for a while, and her son grew up'

The other marker of dependent verbs is the suffix *-andə*, homophonous with the suffixal component of the perfective aspect circumfix *t-...-andə*; consequently, it does not co-occur with tense suffixation. Not surprisingly, given its probable orgin, the usual function of this suffix on dependent verbs is to signal an event in a clause which occurs in a sequential temporal relationship with that of another clause, i.e. event of clause X is complete and then that of clause Y occurs. The dependent clause with the verb marked with *-andə* SEQ can occur either before or after an independent clause:

(7.23) (a) *kwarik ena   mbutunaɲjakosiriandə*
           kwarik ena   mbu-t-naɲja-kosiri-andə
           side   PROX.SG 3.ERG-CAUS-be.at.bank-go.across-SEQ
           nde kwarik abeb   mbutunaɲjakosirionduk
           nde kwarik abeb   mbu-t-naɲja-kosiri-onduk
           so   side   other 3.ERG-CAUS-be.at.bank-go.across-FUT
           'they go across to this side and then will go across to the other side'

   (b) *urukorandə   usirandə   urəmandə*
       u-rukor-andə   u-siri-andə   u-rəmə-andə
       3SG-go.ashore-SEQ   3SG-go.down-SEQ   3SG-stand-SEQ
       mbwerakotandə   Wak yo
       mbu-e-ra-kot-andə   Wak yo
       3.ERG-come-stay-carry-SEQ   PN   DEF
       bubagarikududu
       mbu-mbaŋgari-k-undundu
       3.ERG-kill-FR.PAST-3PC
       'he came ashore and then went down and then stood and then they brought him and they (PC) killed Wak'

   (c) *pate   bibek.   mbukamanakondə*
       pate   mbi-mbi-e-k.   mbu-kam-ana-k-ondə
       before 3DL-AGAIN-come-FR.PAST 3.ERG-give-3SG.DAT-FR.PAST-3DL

        usaranandə
        u-sara-ana-andə
        3SG-report-3SG.DAT-SEQ
        'they (DL) came again soon; they (DL) gave it to her and then he informed her'

(d)   *ayndeb*   *ayndeb*   *mbukamenandə*   *bu-d-andə*
      aynde-mb   aynde-mb   mbu-kame-n-andə   mbu-ndə-andə
      here-OBL   here-OBL   3.ERG-call-LK-SEQ   3.ERG-hear-SEQ
      ŋgiok
      ŋgi-o-k
      3PC-go-FR.PAST
      'they (PC) called out here and there, and then they (PC) heard and then they (PC) went'

(e)   *osenandə*   *aynde*   *pamba*
      o-se-n-andə   aynde   pamba
      go-NIGHT-LK-SEQ   here   just
      ŋgiprəkarorokəto   *nda*   *nor*   *ombe*
      ŋgi-prək-ar-oro-k-to   nda   nor   ombe
      3PL-work-PROG-EXT-FR.PAST-DEP   and   man   INDEF
      [uyarək]   *mbusamaytukondu*
      [u-e-ar-k]   mbu-samayt-k-ondu
      3SG-come-PROG-FR.PAST   3.ERG-see-FR.PAST-3PL
      'they (PL) went at night and then were working here for a while and saw a man coming'

(f)   *mu*   *karondukoya*   *nəmbren*   *mbu-samayt-andə*
      mu   kar-o-ndək-oya   nəmbren   mbu-samayt-andə
      3SG   walk-go-NR.PAST-3SG   pig   3.ERG-see-SEQ
      'he walked about and then saw a pig'

(g)   *nor*   *uren*   *mbutumanəŋangoya*   *onandə*
      nor   uren   mbu-tumanəŋ-ang-oya   o-n-andə
      man   dog   3.ERG-hit-PRES-3SG   go-LK-SEQ
      'the man hits the dog and then went'

(h)   *nor*   *uren*   *mbutumanəŋangoya*   *uren*   *umurandə*
      nor   uren   mbu-tumanəŋ-ang-oya   uren   u-mur-andə
      man   dog   3.ERG-hit-PRES-3SG   dog   3SG-flee-SEQ
      'the man hits the dog and the dog runs away'

Note that the order of the clauses is always iconic according to the unfolding of the events in time. Clauses which describe prior events always precede those which occur later. However, the placement of the suffix -*andə* SEQ is not restricted to verbs in the clauses denoting the prior events; it can just as freely occur on verbs of clauses denoting the subsequent events (7.23c, f, g, h). The suffix simply indicates that there is a sequential relationship between the clauses, its placement being variable, but the actual order of that sequence is determined iconically by the narrated linear order of the clauses.

Unlike with -*to* DEP, clauses marked with -*andə* SEQ can stand by themelves as full utterances, typically in shortish chains of such clauses. This is the phenomenon of insubordination (Evans 2007), often described for other languages:

(7.24) (a) *mbiramanande* e *mbiprenande*
mbi-ra-ma-n-and-e e mbi-pre-n-and-e
3DL-stay-DUR-LK-SEQ-Q or 3DL-die-LK-SEQ-Q
'are they (DL) still alive or are they (DL) dead?'

(b) *ŋgirarorokəŋgaya* *ŋgionandə* Sapend
ŋgi-ra-ar-oro-k-ŋgaya ŋgi-o-n-andə Sapend
3PL-stay-PROG-EXT-FR.PAST-3PL 3PL-go-LK-SEQ PN
yo mayn yowa *ŋgipukoranandə* *tamənd*
yo mayn yowa ŋgi-pukora-ana-andə tamənd
DEF man DIST 3PL-give.food-3SG.DAT-SEQ fish
mak mak yo *ŋgitenandə* *uremayndəpari*
mak mak yo ŋgi-t-e-n-andə ure-mayndəpari
bad bad DEF 3PL-CAUS-come-LK-SEQ dog-man.PL
pukora *ŋga*
pukora ŋga
give.food DAT
'they (PL) stayed for a while and then they went and then they gave food to Sapend, that man, and then they brought the very bad fish to give food to the dog men'

(c) *omanandə* Sumbrakambi *tamənd*
o-ma-n-andə Sumbrakambi tamənd
go-AFTERNOON-LK-SEQ beach fish
uruomanandə *aynde* *pamba*
u-ru-o-ma-andə aynde pamba
3SG-shoot-go-DUR-SEQ here only
'he went in the afternoon to the beach and then was spearing fish just here'

(d) *kadubə   rotandə   gibikisaonandə*
    kandumb   rot-andə   ŋgi-mbi-kisa-o-n-andə
    croton   fall-SEQ   3PC-AGAIN-hide-go-LK-SEQ
    'the croton fell off and they (PC) went to hide again'

It is also possible to combine *-andə* SEQ with *-to* DEP:

(7.25) (a) *kadubə   yo   rotandəto   onandəto*
           kandumb   yo   rot-andə-to   o-n-andə-to
           croton   DEF   fall-SEQ-DEP   go-LK-SEQ-DEP
           'the croton fell off and then they went and'

       (b) *mu   mbutonandə   mu-na   takar   sur.*
           mu   mbu-t-o-n-andə   mu-na   takar   sur.
           3SG   3.ERG-CAUS-go-LK-SEQ   3SG-POSS nest   in
           mbunaŋgromandəto   mbumandəto   karan
           mbu-naŋgro-mə-andə-to   mbu-mə-andə-to   karan
           3.ERG-peck-eat-SEQ-DEP   3.ERG-eat-SEQ-DEP   head
           mbusokamandəto
           mbu-sokam-andə-to
           3.ERG-behead-SEQ-DEP
           'he took him into his nest. He then pecked at him and then he ate him and then beheaded him'

As with many Papuan languages (Foley 2010), the border between clause chaining constructions and serial verb constructions is occasionally somewhat hazy in Kopar. Consider the following example:

(7.26) *indaimbot   mbuturəmakəto*
       inda-imbot   mbu-t-rəmə-k-to
       house-nose   3.ERG-CAUS-stand-FR.PAST-DEP
       mbukinimbəkəto
       mbu-kinimb-k-to
       3.ERG-fasten-FR.PAST-DEP
       'he stood her up and fastened her to the gable and'

Note the 'argument bleeding' so typical of serial verb constructions in (7.26) in spite of the dependent verb marking by *-to* for a clause chaining construction. The noun compound *indaimbot* 'gable' is a semantic argument of the verb *kinimb-* 'fasten', yet it appears in the clause headed by the dependent verb *tu-rəmə-* 'stand up'. This structure makes sense if the sequence of the two verbs forms a kind of complex predicate rather like a serial verb construction taking a single set of arguments which is the sum of their individual arguments. (7.27) is a similar example, this time from the Wongan dialect:

(7.27) ma   tamənd   o   kay   sur   o   rəməkəndi
        ma   tamənd   o   kay   sur   o   rəmə-k-ndi
        1SG   fish   canoe   inside   stand-IRR-ADV
        tənakatiramanandə
        t-na-katira-mana-andə
        PFV-1SG.ERG-spear-DUR-PFV
        'while standing up in a canoe, I had kept on spearing fish'

In this example there is a simultaneous temporal relationship between the two clauses as indicated by the adverbial suffix *-ndi* on the first clause (see section 7.2.1). But note that *tamənd* 'fish', while occuring in the first clause headed by *rəmə-* 'stand' is not an argument of that verb. Rather it is the direct object of the verb of the second clause *katira-* 'spear', yet does not appear in the clause governed by that verb, instead in the preceding one. This is yet another exemplar of 'argument bleeding', indicative of the porous border between serial verb constructions and clause chaining constructions.

Appendix 1

Comparative wordlist of Kopar and Murik

		Kopar	Murik
1.	language	*mimiŋ*	*meneŋ*
2	I	*ma*	*ma*
3	you (SG)	*mi*	*mi*
4.	he/she	*mu/mə*	*mən*
5.	we (DL)	*ke*	*age/gai*
6.	you (DL)	*ko*	*ago/gau*
7.	they (DL)	*mbə*	*məndəb*
8.	we (PC)	*paŋgə*	*agi*
9.	you (PC)	*ŋgu*	*agu*
10.	they (PC)	*məŋgə*	*məŋgə*
11.	we (PL)	*e*	*e*
12.	you (PL)	*o*	*o*
13.	they (PL)	*mbu*	*mwa*
14.	hair on head	*ruar*	*dwar*
15.	head	*karan*	*kambatak/arambatak*
16.	mouth	*asumb*	*sikin*
17.	nose	*imbot*	*ndaur*
18.	eye	*nambrin*	*nabrin*
19.	tongue	*mimiŋ*	*meneŋ*
20.	ear	*kundot*	*karəkep*
21.	tooth	*asirap*	*asarap*
22.	neck	*pətak*	*potak/pwatak*
23.	stomach, belly	*surun*	*sar*
24.	breast	*niŋgin*	*niŋgin*
25.	back	*ruamb*	*dwab*
26.	buttocks	*kandaŋ*	*kayk*
27.	penis	*iriŋ*	*irəg*
28.	liver	*traman*	*tamran*
29.	lung	*tayr*	*pusir*
30.	hand	*napar*	*ndarin*
31.	leg	*naməŋ*	*daŋg*
32.	knee	*pəndəkep*	*danəmbig*
33.	bone	*sarekimb*	*saragip*
34.	blood	*yuaran*	*yaran*
35.	skin	*naŋgun*	*nagun*

36.	person	*nor*	*nor*
37.	male	*pwoyn*	*pwin*
38.	woman, female	*nəma(ndək)*	*nəma(rogo)*
39.	father	*yayan*	*yan*
40.	mother	*nana*	*ŋain*
41.	older parallel sibling	*kakan*	*tatan*
42.	younger parallel sibling	*ram*	*dam*
43.	younger sister of girl	*itəman*	*etəman*
44.	cross sibling	*maman*	*maman*
45.	boy	*ŋaɲjen/naɲjen*	*najen*
46.	girl	*ambiser*	*ŋasen*
47.	bird	*sesen*	*pesen*
48.	dog	*uren*	*oren*
49.	pig	*nəmbren*	*bren/nəmbren*
50.	flying fox	*pəkəp*	*pitak*
51.	wallaby	*yaŋgen*	*yagain*
52.	possum	*inamb*	*enam*
53.	cassowary	*kənd*	*kər*
54.	snake	*ikun*	*wakun*
55.	crocodile	*uri*	*dwamin/oramen/warem*
56.	fish	*taménd*	*tand*
57.	mosquito	*naŋgit*	*nauk*
58.	egg	*awŋ*	*gaug*
59.	water	*arəm*	*arəm*
60.	stone	*iniŋ*	*dug*
61.	fire	*awr*	*aur*
62.	smoke	*atik*	*katik*
63.	sun	*akən*	*akən*
64.	moon	*karep*	*karewan*
65.	star	*kinaŋ*	*birin/mwain*
65.	cloud	*akən-jim*	*pasak*
67.	mountain	*pandam*	*param*
68.	ground	*andin*	*ajin*
69.	road, path	*porakayn*	*yagabor*
70.	tree	*iror*	*yarar*
71.	ironwood tree	*sakənd*	*sakər*
72.	root	*kintip*	*nagəsak*
73.	leaf	*nəmbiraŋ*	*nabirək*
74.	spine of leaf	*kiniŋ*	*kineŋ*
75.	village	*numot*	*nomot*

76.	house	*indan*	*iran*
77.	feast	*kar*	*gar*
78.	meat	*pətak*	*(g)arap*
79.	sago jelly	*numbon*	*bon/numbon*
80.	sago flour	*mareŋ*	*durin*
81.	sugarcane	*muɲamb*	*ajikop*
82.	banana	*piniman*	*peneman*
83.	coconut	*ruaŋ*	*dapag/batak*
84.	betelnut	*puruŋ*	*porog*
85.	betel pepper vine	*aŋətan*	*gatan*
86.	betel powdered lime	*ayr*	*air*
87.	betel refuse	*tak*	*tak*
88.	bow	*mban*	*panain*
89.	spear	*naŋgrin*	*niŋgek*
90.	axe	*panden*	*borin/emən*
91.	canoe	*kayn*	*gayn*
92.	oar	*inaŋ*	*inaŋg*
93.	basket	*sapwar*	*sun/sumon*
94.	net bag	*maɲjipak*	*marep*
95.	hearth	*aymbor*	*nabren*
96.	mosquite net	*andəŋ*	*arəg*
97.	drum (garamut)	*nəmbən*	*dəbən*
98.	hourglass drum (kundu)	*nigedip*	*wagən*
99.	bamboo flute	*rikam*	*dəkab/bawr*
100.	male cult house	*andimeŋ*	*tab*
101.	spirit	*paraŋ*	*brag*
102.	thing	*moran*	*moran*
103.	who (SG)	*menome*	*menamena*
104.	what	*ara(nomena)*	*amanamena*
105.	where	*ndokome*	*magaŋamena*
106.	one	*ombe/mbiona/mbatep*	*abe*
107.	two	*kombar(i)*	*kobo*
108.	three	*keremən*	*keroŋgo*
109.	four	*sanandək*	*saŋaŋgo*
110.	five	*tambək*	*tabugo*
111.	six	*tambək mbatepanda*	*(tabugo) batep ave*
112.	seven	*tambək koɲjiranda*	*(tabugo) batep kobo*
113.	eight	*tambək keremənganda*	*(tabugo) batep keroŋgo*
114.	ten	*aytapor*	*darabor (abe)*
115.	twenty	*pwoyn mbatep*	*darabor kobo*

116.	yes	*awo*	*awo*
117.	morning	*tumbuna*	*patebaba*
118.	afternoon	*wakənan*	*akən nakomb*
119.	night	*rakamdan*	*dagam*
119.	before	*patemba*	*pateba*
120.	today	*ndesa*	*kote*
121.	yesterday/tomorrow	*rari*	*ŋarən*
122.	two days before/after	*(rari) uta*	*utaga*
123.	big	*kapu*	*apo*
124.	small	*ŋaɲjirik*	*gaŋgas*
125.	good	*aratək*	*arato*
126.	bad	*mak*	*mwago*
127.	white	*kaymbak*	*kakrep*
128.	black	*petəndək*	*nuŋgun/pasigan*
129.	long	*kuŋgoparik*	*kogoŋgo*
130.	short	*katarək*	*pokəp*
131.	heavy	*patəndək*	*dipatogo*
132.	light, not heavy	*bijabətək*	*sanabran*
133.	old	*patendək*	*patero*
134.	new	*nəŋgəmək*	*deŋarero*
135.	hot	*ududuk*	*awuraro*
136.	cold	*sarapakin*	*kokobet*
137.	many, much	*awtok*	*abetabeta*
138.	above	*kaŋgarap*	*araŋap*
139.	below	*tawmb/tomb*	*taumb/tomb*
140.	behind	*tiŋgi*	*kaego*
141.	close	*asama(ri)*	*pasamari*
142.	far	*patuku(ri)*	*iɲji*
143.	inside	*sur*	*sar*
144.	middle	*pwap*	*pwap*
145.	hunger	*nime*	*nəmre*
146.	shame	*iɲarapin*	*yaŋarib/yaŋabin*
147.	smell	*munəmb*	*munəmb*
148.	mind, thought	*nunon*	*nonon*
149.	see	*samayt-/sobo-*	*sobo-*

The second Kopar form *sobo-* is very likely a Murik loan, as it has a plain voiced stop, which otherwise only occurs in Kopar through denasalization from a prenasalized voiced stop in an adjoining syllable.

150.	hear	ndə-	də-
151.	know	ndat-	dat-
152.	become, feel, do	si-	si-
153.	stay, be at	ra-	da-
154.	sit	ndasa-	sasa-
155.	stand	rəmə-	dəm-
156.	lie, sleep	kandək-	ara-
157.	die	pre-	pre-
158.	urinate	miɲjir pra-	miɲjir pra-
159.	defecate	məndən pra-	məndən pra-
160.	give birth	mbe-	be-
161.	walk	kar-	ar-
162.	come	e- ~ ya-	e- ~ ya-
163.	go	o- ~ wa-	o-
164.	bring	t-e-	t-e-
165.	take	t-o-	t-o-
166.	leave	sambo-	sabwa-
167.	get up	riker-	dekara-
168.	wait	samer-	samer-
169.	go down	siri-	sir-
170.	fall	rot-	dot-
171.	go ashore	rukor-	dokor-
172.	build	undi-	uji-
173.	eat, drink	mə-	mə-
174.	feed	pukora-	pakor-
175.	pound sago	punu-	pon-
176.	wash sago	tuku-	tuku-
177.	chew betelnut	kanam-	anam-
178.	give	kam-	aməŋ-
179.	buy	uki-	oki-
180.	try	tay-	tei-
181.	open	pwa-	pwa-
182.	fasten	kinimb-	kimb-
183.	call out	kamen-	amen-/tamen-
184.	cough	rukwa-	dokwa-
185.	shoot	ru-	ur-
186.	here	aynde	ane
187.	this (SG) dog	uren enana	oren ewa
188.	that dog	uren yowa	oren dewa

189.	my dog	ma-na uren	ma-na oren
		1SG-POSS dog	1SG-POSS dog
190.	I slept	tə-ma-kandək-a	tə-man-ar-a
		PFV-1SG.NOM-sleep-PFV	PFV-1SG.NOM-sleep-TNS
191.	he slept	tə-Ø-kandək-a	t-o-ar-a
		PFV-3SG.NOM-sleep-PFV	PFV-3SG.NOM-sleep-TNS
192.	I'm standing	ma-rəm-aŋg-aya	mana-dəm-ar-a
		1SG.NOM-stand-PRES-1SG	1SG.NOM-stand-PROG-TNS
192.	he is eating	u-m-ar-aŋg-oya	o-m-ar-a
		3SG.NOM-eat-PROG-PRES-3SG	3SG.NOM-eat-PROG-TNS
194.	I ate	tə-ma-m-a	tə-mana-mə
		PFV-1SG.NOM-eat-1SG.PFV	PFV-1SG.NOM-eat
195.	I saw him	na-samayt-nduk-oya	o-a-sobo-ra
		1SG.ERG-see-PAST-1SG	3SG.NOM-1SG.ERG-see-TNS
196.	he saw me	tə-ŋga-samayt-a	tə-n-aŋa-sobo-ŋa
		PFV-INV-see-1SG.PFV	PFV-INV-1SG.ACC-see-INV
197.	he saw him	tu-mbu-samayt-a	t-o-sobo-ra
		PFV-3.ERG-see-3SG.PFV	PFV-3SG.NOM-see-TNS

198. he gave me betelnut

    Kopar: puruŋ ma ŋga mbu-kam-unduk-oya or
             betelnut 1SG DAT 3.ERG-give-PAST-3SG

            puruŋ mbu-kam-əndək-ənaya
            betelnut 3.ERG-give-PAST-1SG.NOM

    Murik: porog o-aŋa-aməŋ-ana
             betelnut 3SG.NOM-1SG.ACC-give-TNS

199. I feel ashamed

    Kopar: ma iŋarapin t-Ø-si-a
             1SG shame PFV-3SG-do-PERF

    Murik: ma yaŋarib t-o-aŋa-ri-ŋa
             1SG shame PFV-3SG.NOM-1SG.ACC-shoot-INV

200. three woman are eating

    Kopar: nəmandəpak keremən mora ŋgi-m-ar-aŋg-iya
             women.PL three thing 3PC/PL.NOM-eat-PROG-PRES-PC

    Murik: nəmaragara keroŋgo mora də-m-ar-a
             woman.PC three thing 3PC.NOM-eat-PROG-TNS

Appendix 2

Text: irormandək   nda   muna   ambisen   nda
       iror-mandək nda   mu-na   ambisen   nda
       tree-female and   3SG-POSS   daughter and
       'a tree spirit and her daughter'

Story narrated by Paul Mason of Kopar village in January 1995. Recording and initial transcription by Paul Mason and Stephen Hill. Final transcription by me.

1. mbu   patemba   nəmandəŋ   urəm   ŋgiosirikəŋgaya
   mbu   pate-mba   nəmandək   urəm   ŋgi-o-siri-k-ŋgaya
   3PL   before-LOC   woman   kin.group 3PL-go-go.down-FR.PAST-3PL
   kumoy   ŋgarin
   kumoy   ŋgari
   fish.with.net   DAT
   'a long time ago a kin group of women went down to fish with nets'

2. mbə   nanan   nda   ambisen   nda   kayn   mbatep
   mbə   nana   nda   ambisen   nda   kayn   mbatep
   3DL   mama   and   daughter and   canoe   one
   mbitəkandaŋakindi,   mbiosirik
   mbi-t-kanda-ŋa-k-ndi   mbi-o-siri-k
   3DL-COM-board-go.into-FR.PAST-ADV   3DL-go-go.down-FR.PAST
   'a mama and daughter jumped together into one canoe and went downriver'

3. mbiosirikəmbaya,   pwapindi   ambisen   surun   nda
   mbi-o-siri-k-mbaya   pwap-ndi   ambisen   surun   nda
   3DL-go-go.down-FR.PAST-3DL   middle-ADV daughter belly   COM
   usianak
   u-si-ana-k
   3SG-do-3SG.DAT-FR.PAST
   'they (DL) both went down river and midway the daughter felt pregnant'
   (literally 'feels with her belly', a gloss on the Tok Pisin expression *i gat bel* for 'pregnant')

4. nanan   nda   mbutəmek   'nana,   ma   mambon
   nana   nda   mbu-təme-k   nana   ma   mambon
   mama   COM   3.ERG-tell-FR.PAST   mama   1SG   spine

	təsya	aynde.	ŋgakaketamənda
	t-Ø-si-a	aynde	ŋga-ka-ketam-nda
	PFV-3SG-do-PFV	here	INV-FIRST-put.ashore-IMP.1.OBJ

mandakamben
ma-nda-ka-mbe-n
ITR.IMP-NOW-FIRST-bear-1SG

'she told her mama 'mama I feel my spine here; bring me ashore first, and let me now give birth first now''

5. nda nana   mbuturukorukəto,   pwapindi
   nda nana   mbu-t-rukor-k-to   pwap-ndi
   and mama   3.ERG-CAUS-go.ashore-FR.PAST-DEP   middle-ADV
   mbuketaməkə,   mbusambotək
   mbu-ketam-k   mbu-sambo-t-k
   3.ERG-put.ashore-FR.PAST 3.ERG-leave-APPL-FR.PAST

   'and mama brought her to the river bank and put her ashore in the middle (of the bank) and left her there'

6. nda nanan onak   ukumoyomanak
   nda nanan o-na-k   u-kumoy-o-ma-na-k
   and mama go-UP-FR.PAST   3SG-fish.with.net-go-DUR-UP-FR.PAST

   'and mama went upriver, she went upriver fishing with a net as she went'

7. ukumoywarəkə,   nda ambisen aynde
   u-kumoy-o-ar-k   nda ambisen aynde
   3SG-fish.with.net-go-PROG-FR.PAST and daughter here
   uramanakəto,   umbekəto   semayndək,
   u-ra-ma-na-k-to   u-mbe-k-to   sen-mayndək
   3SG-stay-DUR-?-FR.PAST-DEP 3SG-bear-FR.PAST-DEP son-male
   nana   ŋga   usamerəmak
   nana   ŋga   u-samer-ma-k
   mama   DAT   3SG-wait-DUR-FR.PAST

   'while she was going along fishing with nets, the daughter remained here and gave birth to a son and waited for her mama'

8. nda mbu gibiuyamanyarək
   nda mbu ŋgi-mbi-uyaman-e-ar-k
   and 3PL 3PL-AGAIN-turn.back-come-DAY-FR.PAST

   'and they (PL) (the original kin group of women) turned back and were coming back during daytime'

9. ŋgiyarəkəŋgaya,   kayŋgaran   ŋgara etayuk
    ŋgi-e-ar-k-ŋgaya   kay-ŋgaran   ŋgara etayu-k
    3PL-come-DAY-FR.PAST-3PL   canoe-front   first approach-FR.PAST
    yowa   nda   ambise   enana   mbukararatak
    yowa   nda   ambise   enana   mbu-karara-ta-k
    DIST   and   daughter PROX.SG   3.ERG-ask-OUT-FR.PAST
    'They (PL) came and this daughter asked the lead canoe which came first'

10. '*mana   nana   o'   nda   mbutəmekududu*
    ma-na   nana   o   nda   mbu-təme-k-undundu
    1SG-POSS   mama   oh   and   3.ERG-say-FR.PAST-3PC
    'my mother, o?' and they (PC) told her'

11. '*mina   nanan   ɲja   muna   tiŋgi   o'*
    mi-na   nanan   ɲja   mu-na   tiŋgi   o
    2SG-POSS   mama   still 3SG-POSS   behind   oh
    'your (SG) mama is still behind'

12. *aynde   pamba   uramakəto,*   *nanan   nuŋgo*
    aynde   pamba   u-ra-ma-k-to   nanan   nuŋgo
    here   only   3SG-stay-AFTERNOON-FR.PAST-DEP   mama   very
    tiŋgi   o   emanak
    tiŋgi   o   e-ma-ana-k
    behind oh come-AFTERNOON-3SG.DAT-FR.PAST
    'she remained only there until afternoon and mama was coming to her very late in the afternoon'

13. *enakoya,*   *asama   enakoya,*
    e-ana-k-oya   asama   e-ana-k-oya
    come-3SG.DAT-FR.PAST-3SG   close   come-3SG.DAT-FR.PAST-3SG
    mbukamek
    mbu-kame-k
    3.ERG-call.out-FR.PAST
    'she came toward her and (when) she came close to her, she (the daughter) called out'

14. *'nana   nda   maenato*
    nana   nda   ma-e-ona-to
    mama   and   ITR.IMP-come-2SG-DEP
    ma   gadaerakotəndi   ya
    ma   ŋga-nda-e-ra-kot-nd   ya
    1SG   INV-NOW-come-stay-carry-IMP.1.OBJ   EMPH
    'mama, come and get me'

15. *nanan   urikeranakoya   ukatanak*
    nanan   u-riker-ana-k-oya   u-kata-ana-k
    mama   3SG-get.up-3SG.DAT-FR.PAST-3SG 3SG-speak-3SG.DAT-FR.PAST
    'mama stood up toward her and said to her'

16. *'awo, ma   mi   mbe   ŋga   mbutenaŋgənaya*
    awo   ma   mi   mbe   ŋga   mbu-t-e-n-aŋg-naya
    yes   1SG   2SG   bear   DAT   3.ERG=1-CAUS-come-LK-PRES-2SG
    'yes, I brought you (SG) to give birth'

17. *mi   pətak   erakot   ŋga   ekəna   mbasi*
    mi   pətak   e-ra-kot   ŋga   e-k-na   mbasi
    2SG   meat   come-stay-carry   DAT   come-IRR-2SG COUNTERFACT?
    'but I thought you (SG) came to get meat (protein food, Tok Pisin *abus*)'

18. *aynde   maramaka,   məndən   kumuka   miɲjir*
    aynde   ma-ra-ma-ka   məndən   ku-mu-ka   miɲjir
    here   ITR.IMP-stay-DUR-2SG   feces   TR.IMP-eat-2SG urine
    kumuka   mina   naɲjenak   mapreka!'
    ku-mu-ka   mi-na   naɲjen-na-k   ma-pre-ka
    TR.IMP-drink-2SG   2SG-POSS   child-POSS-NE   ITR.IMP-die-2SG
    'you (SG) stay here, eat the feces and drink the urine of your (SG) child and die!'

19. *'ee,   nana,   sapikindina,   moka   kena   ŋga*
    'ee   nana   sapiki-ndi-ona   ma-o-ka   ke-na   ŋga
    ok,   mama   right-ADV-2SG   ITR.IMP-2SG 1DL-POSS   DAT
    ŋasambotənda
    ŋga-sambo-t-nda
    INV-leave-APPL-IMP.1.OBJ
    'eh, mama, you (SG) are right, you (SG) go and leave us (DL)'

20. *uramak,*   *ambisen uramakə,*   *arəm*
    u-ra-ma-k   ambisen   u-ra-ma-k   arəm
    3SG-stay-DUR-FR.PAST daughter 3SG-stay-DUR-FR.PAST water
    idiyaranak
    idi-e-ar-ana-k
    rise?-come-PROG-3SG.DAT-FR.PAST
    'she stayed, the daughter stayed, and the river began rising on her'

21. *arəm   idienakə,*   *namən*
    arəm   idi-e-ana-k,   namən
    water   rise?-come-3SG.DAT-FR.PAST leg
    mbwarakotanak
    mbu-wa-ra-kot-ana-k
    3.ERG-go-stay-carry-3SG.DAT-FR.PAST
    'the river rose on her and it reached her leg'

22. *namən o ndək eranakə,*   *mambon*
    namən   o ondək e-ra-ana-kə,   mambon
    leg   PURP   come-stay-3SG.DAT-FR.PAST spine
    mbwarakotanak
    mbu-wa-ra-kot-ana-k
    3.ERG-go-stay-carry-3SG.DAT-FR.PAST
    'coming up from her leg, it reached her back'

23. *ŋaɲjen mbuturikerək*   *mbukiketək*
    ŋaɲjen   mbu-t-riker-k   mbu-kiket-k
    child   3.ERG-CAUS-get.up-FR.PAST 3.ERG-hold.under.arm-FR.PAST
    'she lifted up her child and held him under her arm'

24. *mambondək   eranakə,*   *rakon*
    mambon-ndək   e-ra-ana-k   rakon
    spine-PURP   come-stay-3SG.DAT-FR.PAST shoulder
    mbuwarakotanak   *ŋaɲjen*
    mbu-wa-ra-kot-ana-k   ŋaɲjen
    3.ERG-go-stay-carry-3SG.DAT-FR.PAST   child
    mbumoɲjak
    mbu-moɲja-k
    3.ERG-hold.on.shoulder-FR.PAST
    'coming up from her back, it reached her shoulder and she put her child on her shoulders'

25. *rakondək   eranakə,   karanpi   ŋgari   ŋaɲjen*
    rakon-ndək   e-ra-ana-k   karan-pi ŋgari   naɲjen
    shoulder-PURP   come-stay-3SG.DAT-FR.PAST   head-just DAT   child
    mbuturikerək   kara   kaŋgarap
    mbu-t-riker-k   kara   kaŋgarap
    3.ERG-CAUS-get.up-FR.PAST   head   above
    'coming up from her shoulder, right up to her head, she lifted up her child above her head'

26. *irormandək   usiranak   naɲjen*
    iror-mandək u-siri-ana-k   naɲjen
    tree-female 3SG-go.down-3SG.DAT-FR.PAST   child
    mbwerakotanak
    mbu-e-ra-kot-ana-k
    3.ERG-come-stay-carry-3SG.DAT-FR.PAST
    'a female tree spirit came down to her and seized her child'

27. *'sen   kute   mi   mana   ambise   a*
    sen   ku-t-e   mi   ma-na   ambise   a
    son   TR.IMP-CAUS-come   2SG   1SG-POSS daughter
    marukorəka   mandaoke   mana   inda   sur'
    ma-rukor-ka   ma-nda-o-ke   ma-na   inda   sur'
    ITR.IMP-go.ashore-2SG   ITR.IMP-NOW-go-1DL   1SG-POSS house into
    'give (me) a son, you (SG) are my daughter, you (SG) come ashore and let us (DL) go to my house'

28. *irormandək   enana   mbətokə   muna   inda*
    iror-mandək   enana   mbu-t-o-k   mu-na   inda
    tree-female   PROX.SG   3.ERG-CAUS-go-FR.PAST   3SG-POSS   house
    sur   o, mbutəmek
    sur   o   mbu-təme-k
    into   3.ERG-tell-FR.PAST
    'this female tree spirit took her to her house and told her'

29. *'murarəkənaya   mana   ŋgari,   mana   ambise   mina*
    mur-arək-naya   ma-na   ŋgari   ma-na   ambise   mi-na
    fear-PROHIB-2SG   1SG-POSS DAT   1SG-POSS   daughter   2SG-POSS

ka-kerondək   o   uprekoya
ka-kerondək   o   u-pre-k-oya
same-kind   3SG-die-FR.PAST-3SG
'don't be afraid of me, my daughter who was the same as you (SG) died'

30. ma   ambise   muarim   yuwaya   natəraomaŋgara
    ma   ambise   muarim   yuwa-oya   na-t-ra-o-ma-aŋg-ara
    1SG   daughter grief   have-1SG   1SG.ERG-COM-stay-go-DUR-PRES-?
    nda   mi   nda   mana   ambise   simbina'
    nda   mi   nda   ma-na   ambise   si-mbi-ona
    and   2SG   COM   1SG-POSS   daughter   become-IM.FUT-2SG
    'I have grief over my daughter, I stay with you (SG), and you (SG) will become my daughter'

31. mbuturukorək   mumoran   mbusianak
    mbu-t-rukor-k   mə-moran   mbu-si-ana-k
    3.ERG-CAUS-go.ashore-FR.PAST   eat-thing   3.ERG-do-3SG.DAT-FR.PAST
    mumoran   mbupukorak
    mə-moran   mbu-pukora-k
    eat-thing   3.ERG-serve.food-FR.PAST
    'She took her ashore and made her food and served her food'

32. nda   nəmandək   raway   ambise   muna   ŋga
    nda   nəmandək   raway   daughter   mu-na   ŋga
    and   woman   self   daughter   3SG-POSS   DAT
    ukatanak
    u-kata-ana-k
    3SG-speak-3SG.DAT-FR.PAST
    'and the woman herself, the daughter, said to her'

33. 'nana,   indan   mina   ndokomena?'
    nana   indan   mi-na   ndok-omena
    mama   house   2SG-POSS   where-wh
    'mama, where is your (SG) house?'

34. 'indan   minak   yowana   mo   irorinda   mararaŋənaya'
    indan   mi-na-k   yowa-ona mo   iror-inda   ma-ra-ar-aŋg-naya
    house   2SG-POSS-NE   DIST-2SG ?   tree-house   1SG-stay-PROG-PRES-1SG
    'you (SG) have that house of yours (SG). I live in a tree house'

35. rakamda ŋgikandəksekəto,   nanan
    rakamda ŋgi-kandək-se-k-to   nanan
    night   3PC-sleep-NIGHT-FR.PAST-DEP mama
    urikerəkəto,   indan kapu suman o
    u-riker-k-to   indan kapu suman o
    3SG-get.up-FR.PAST-DEP   house big   very
    ŋgosi indan mbuturikeranak
    ŋgosi indan mbu-t-riker-ana-k
    real house 3.ERG-CAUS-get.up-3SG.DAT-FR.PAST
    'they (PC) slept during the night, and mama got up and erected a very big house, a real house, for her'

36. nda ambisen ukandəksenakoya   miɲjir
    nda ambisen u-kandək-se-ana-k-oya   miɲjir
    and daughter 3SG-sleep-NIGHT-3SG.DAT-FR.PAST-3SG urine
    pra   ŋga usianakoya
    pra   ŋga u-si-ana-k-oya
    excrete DAT 3SG-do-3SG.DAT-FR.PAST-3SG
    utuməranəmanakoya   miɲjir
    u-təməran-ma-ana-k-oya   miɲjir
    3SG-roll.over-DUR-3SG.DAT-FR.PAST-3SG urine
    upranakoya   manduwara nəmbiraŋ o
    u-pra-ana-k-oya   manduwara nəmbiraŋ o
    3SG-excrete-3SG.DAT-FR.PAST-3SG taro   leaf   O
    piniman imbian sur o miɲjir urotanak
    piniman imbian sur o miɲjir u-rot-ana-k
    banana leaf   onto   urine 3SG-fall-3SG.DAT-FR.PAST
    miɲjir bubitəkatamək
    miɲjir bu-bi-t-katam-k
    urine   3.ERG-AGAIN-CAUS-noise-FR.PAST
    'while her daughter slept during the night, she felt like urinating, and she rolled over and urinated, and her urine fell on taro leaves and banana leaves so that the urine made a noise again'

37. 'ma ndime! nana inda ombe tumbusya'
    ma ndi-ome nana inda ombe t-mbu-si-a
    1SG how-wh mama house a   PFV-3.ERG-do-PFV
    'how about me! mama built a house'

38. nambrin   mbutupuasaməkoya   indan   kapu   suman
    nambrin   mbu-t-puas-am-k-oya   indan   kapu   suman
    eye   3ERG-CAUS-open-DETR-FR.PAST-3SG   house   big   very
    mbusamaytək
    mbu-samayt-k
    3.ERG-see-FR.PAST
    'she opened her eyes and saw a very big house'

39. akən   upratanakoya   nda   ambisen
    akən   u-pra-ta-ana-k-oya   nda   ambisen
    sun   3SG-excrete-OUT-3SG.DAT-FR.PAST-3SG   and   daughter
    mbutəmek
    mbu-təme-k
    3SG-tell-FR.PAST
    'the sun came up on her and she (the female tree spirit) told her daughter'

40. 'ma   nana   nəmandək   endekerondəkənaya'
    ma   nana   nəmandək   ende-kerondək-naya
    1SG   mama   woman   here-kind-1SG
    'I am a mama, I am this kind of woman'

41. 'indan   aratək   mana   ŋga   tumbuturikera'
    indan   aratək   ma-na   ŋga   t-mbu-t-riker-a
    house   good   1SG-POSS   DAT   PFV-3.ERG=2/1.OBJ-CAUS-get.up-PFV
    'you (SG) erected a good house for me'

42. ambisen   aymbor   kwarək   abeb   nanan   aymbor   kwarək   abeb
    ambisen   aymbor   kwarək   abeb   nanan   aymbor   kwarək   abeb
    daughter   stove   side   one   mama   stove   side   one
    nana   nda   aynde   ŋgirarorokəto   sen   kapu
    nana   nda   aynde   ŋgi-ra-ar-oro-k-to,   sen   kapu
    mama   COM   here   3PC-stay-PROG-EXT-FR.PAST-DEP   son   big
    usianak
    u-si-ana-k
    3SG-do-3SG.DAT-FR.PAST
    'the daughter with her stove on one side, mama with her stove on another side, they stayed with mama for a long time, and her son grew up'

43. nanan yowa pətak erasowarənda nimbren o tamənd o
    nanan yowa pətak eras-o-ar-nda nimbren o tamənd o
    mama DIST meat search-go-PROG-COM pig fish
    mbutukasayaroronak muna ŋga mə
    mbu-t-kasay-ar-oro-ana-k mu-na ŋga mə
    3.ERG-CAUS-heap-PROG-EXT-3SG.DAT-FR.PAST 3SG-POSS DAT eat
    ŋga rumbuna enamb
    ŋga rumbuna ena-mb
    DAT grandchild PROX.SG-OBL
    'this mama went looking for fish, she kept on heaping up pork and fish for her grandson to eat'

44. nda rumbunan kapu usianakəto, nor kapu
    nda rumbunan kapu u-si-ana-k-to nor kapu
    and grandchild big 3SG-do-3SG.DAT-FR.PAST-DEP man big
    usianak o nda e nda usianak
    u-si-ana-k o nda e nda u-si-ana-k
    3SG-do-3SG.DAT-FR.P go and come and 3SG-do-3SG.DAT-FR.PAST
    mban ŋga urarianak
    mban ŋga u-rari-ana-k
    bow DAT 3SG-cry-3SG.DAT-FR.PAST
    'and her grandchild having grown up, as an adult, he liked coming and going, and he cried to her for a bow'

45. muna apase enana irormandək mban
    mu-na apase enana iror-mandək mban
    3SG-POSS grandparent PROX.SG tree-female bow
    mbusianak nandada
    mbu-si-ana-k nanda-nda
    3.ERG-make-3SG.DAT-FR.PAST arrow-COM
    'his grandmother, the female tree spirit, made a bow for him with arrows'

46. nda mu Sumbrakambi aynde uruomanandə,
    nda mu Sumbrakambi aynde u-ru-o-ma-n-andə
    and 3SG place.name.at.beach here 3SG-shoot-go-DUR-LK-SEQ
    sese rarisak enambo piniman mə nda
    sese rarisak ena-mbo piniman mə nda
    bird small.bird.species PROX.SG-OBL banana eat COM

oma   yowa
o-ma   yowa
go-DUR DIST

'and he went around shooting at the beach Sumbrakambi for small birds, which were going there eating bananas'

47. surukur   o   men   o   taməndiŋ   okumbi
    surukur   o   men   o   taməndiŋ   okumbi
    kookaburra   sparrow   bird.species PL
    mburumondiat   omanandə   kaykor   o
    mbu-ru-mondi-at   o-ma-n-andə   kaykor   o
    3.ERG-shoot-penetrate-?   go-DUR-LK-SEQ   white.cockatoo
    ukusamandə   Sumbrakambi   tamənd
    u-kusa-am-andə   Sumbrakambi   tamənd
    3SG-go.out-DETR-SEQ   place.name.at.beach   fish
    uruomanandə
    u-ru-o-ma-n-andə
    3SG-shoot-go-DUR-LK-SEQ

'he kept on shooting birds as he went, kookaburras, sparrows, white cockatoos etc; he came out onto the beach Sumbrakambi and went along shooting fish'

48. aynde pamba aken abeb   apasen   yowa onakoya
    aynde pamba aken abeb   apasen   yowa o-na-k-oya
    here   only   sun   another   grandparent DIST go-UP-FR.PAST-3SG

'just here on another day, that grandmother went upriver'

49. onakoya   səmbər nda yan   kombari ram
    o-na-k-oya   səmbər nda yan   kombari ram
    go-UP-FR.PAST-3SG   forest and papa two   younger.brother
    kakan   nda mbə naŋatik   tamənd
    kakan   nda mbə naŋatik   tamənd
    older.brother and 3DL river.current fish
    mbiruesirik
    mbi-ru-e-siri-k
    3DL-shoot-come-go.down-FR.PAST

'she went upriver into the forest and two men, a younger and an older brother, were shooting fish on the current as they came downriver'

50. mbiesirikəto   nəma   sur aynde sen
    mbi-e-siri-k-to   nəma   sur aynde sen
    3DL-come-go.down-FR.PAST-DEP middle.of.river in   here son
    yo   nambrin onakoya   mbusamayndakoya
    yo   nambrin o-ana-k-oya   mbu-samay-anda-k-oya
    DEF eye   go-3SG.DAT-FR.PAST-3SG 3.ERG-see-3DL.DAT-FR.PAST-3SG
    nda raposərukorək   inda sur o nanan mbutəmek
    nda rapos-rukor-k   inda sur o nanan mbu-təme-k
    and run-go.ashore-FR.PAST house into o mama   3.ERG-tell-FR.PAST
    'they (DL) came downriver, and here in the middle of the river, the son
    looked up and saw them (DL) and he ran and went ashore into the house
    and told his mama'

51. 'nana, nor kombari mbe embaya'
    mama   man two   ?   PROX.DL
    'mama, there are these two men'

52. nana   ukusamb   mbusamaytəkəto   katanak
    nana   u-kusa-am   mbu-samayt-k-to   kata-ana-k
    mama   3SG-go.out-DETR 3.ERG-see-FR.PAST-DEP speak-3SG.DAT-FR.PAST
    'mama went out and saw them (DL) and said to him'

53. 'yan o   mbam o   mina   yan   ukataesirande?'
    yan   o   mbam o   mi-na   yan   u-kata-e-siri-andə-e
    papa maybe   2SG-POSS   papa 3SG-talk-come-go.down-PFV-Q
    'I think maybe your papa has come?'

54. mbə mbirikerək   simariŋ   iror su   mbipukak
    mbə mbi-riker-k   simariŋ   iror su   mbi-puka-k
    3DL 3DL-get.up-FR.PAST   tree.species tree onto 3DL-climb-FR.PAST
    'they (DL) got up and climbed a simariŋ tree'

55. mbiokə,   kəmbək   mbindasakəto,   nambrin o
    mbi-o-k   kəmbək   mbi-ndasa-k-to   nambrin o
    3DL-go-FR.PAST   tree.crown 3DL-sit-FR.PAST-DEP   eye
    omba usirsiranəmbak
    omba u-siri-siri-anəmba-k
    ?   3SG-go.down (siri- RED)-3DL.DAT-FR.PAST
    'they (DL) went and sat down in the crown of the tree, and they (DL) watched
    them (DL) (literally 'the eye went down on them (DL)')

56. nda   yan   muna   ramandək   nda   enakoya
    nda   yan   mu-na   ramandək   nda   e-na-k-oya
    and   papa   3SG-POSS younger.brother   COM   come-UP-FR.PAST-3SG
    asamari   nda   mbutəmek   sen
    asamari   nda   mbu-təme-k   sen
    close   and   3.ERG-tell-FR.PAST   son
    'and the papa with his younger brother came up close and she told her son'

57. 'yan   ena   yan   komba   mbaysirangəmbaya,   yan
    yan   enana   yan   kombari   mbi-ay-siri-ang-mbaya   yan
    papa   PROX.SG   papa   two   3DL-?-go.down-PRES-3DL papa
    minak   munak   rama   nda
    mi-na-k   mu-na-k   rama   nda
    2SG-POSS-NE   3SG-POSS-NE   younger.brother   COM
    naysirangoya'
    na-ay-siri-ang-oya
    3SG-?-go.down-PRES-3SG
    'this papa, the two papas are coming downriver; your (SG) papa with his younger brother is coming downriver'

58. nda   aynde   mbirasirikəto,   mbə   mbiekə
    nda   aynde   mbi-ra-siri-k-to   mbə   mbi-e-k
    and   here   3DL-stay-go.down-FR.PAST-DEP   3DL   3DL-come-FR.PAST
    sambariŋ   tabragay,   mbə   sambariŋ   kambur   kombari
    sambariŋ   tabragay   mbə   sambariŋ   kambur   kombari
    tree.species   underneath   3DL   tree.species   shoot   two
    bubierakotəkondə   pinimb   enamb
    mbu-mbi-e-ra-kot-k-ondə   pinimb   ena-mb
    3.ERG-AGAIN-come-stay-carry-FR.PAST-3DL   finger   PROX.SG-OBL
    mbu-kak-ondə   ombe   mbukakawarumbutukondə
    mbu-ka-k-ondə   ombe   mbu-ka-kawarumbut-k-ondə
    3.ERG-cut.into-FR.PAST-3DL   one   3.ERG-FIRST-throw.down-FR.PAST-3DL
    ŋgara   ambe   mbe   bubikawarumbutukondə
    ŋgara   ambe   mbe   bu-bi-kawarumbut-k-ondə
    first   another   one?   3.ERG-AGAIN-throw.down-FR.PAST-3DL
    'and while they (DL) (mama and son) stayed here, they (DL) (two brothers) came to underneath the *sambariŋ* tree (where the mama and son are sitting in the crown), and they (DL) (mama and son) took two shoots of the *sambariŋ* tree and pierced them with a finger and threw one down first and then they (DL) again threw another one down'

59. mayn   ramay   naɲjirik ukatanak
    mayn   ramay   naɲjirik u-kata-ana-k
    husband   younger.brother   little   3SG-speak-3SG.DAT-FR.PAST
    'the husband (e.g. the older brother) said to the younger little brother'

60. 'e,   sambariŋ   kambur   yowa-na   inamb   o   mbam   o'
    e   sambariŋ   kambur   yowa-ona   inamb   o   mbam   o
    hey   tree.species   shoot   DIST-2SG   possum   maybe
    'hey, you (SG) have those *sambariŋ* shoots, maybe a possum'

61. nda   mayn   yowa   nimbep   mbwerakotanak
    nda   mayn   yowa   nimbep   mbu-e-ra-kot-ana-k
    and   husband DIST   spear   3.ERG-come-stay-carry-3SG.DAT-FR.PAST
    nda   mu   nda   nimbep   mbutupapaneɲjakəto
    nda   mu   nda   nimbep   mbu-t-papaneɲja-k-to
    and   3SG   and   spear   3.ERG-CAUS-be.aimed.RED-FR.PAST-DEP
    nda   mbutəmek   'ndokomena?'
    nda   mbu-təme-k   ndok-omena
    and   3.ERG-tell-FR.PAST   where-wh
    'and that husband got his spear and he aimed his spear and told him 'where?''

62. nda   mbə   mbikinaŋgamək
    nda   mbə   mbi-kinaŋ-am-k
    and   3DL   3DL-appear=star-DETR-FR.PAST
    'and they (DL) (mama and son) appeared'

63. 'kona   inamb   o   ko   ma   kuŋganambratəmbiko'
    ko-na   inamb   o   ko   ma   ku-ŋga-nambrat-mbi-ko
    2DL-POSS   possum   2DL   1SG   TR.IMP-FIRST-spear-IM.FUT-2DL
    '(this is) your (DL) possum, you (DL) spear me now'

64. nda   mayn   yowa   nor   ɲaɲjirik   enana   yan   yo
    nda   mayn   yowa   nor   ɲaɲjirik   ena-na   yan   yo
    and   husband DIST   man   little   PROX.SG-POSS   papa   DEF
    nda   nimbep   napar   sur   o   ndək   urotanak
    nda   nimbep   napar   sur   o   ndək   u-rot-ana-k
    and   spear   hand   inside   PURP   3SG-fall-3SG.DAT-FR.PAST

nda urarianak
nda u-rari-ana-k
and 3SG-cry-3SG.DAT-FR.PAST

'and the husband, the father of this little man (the son), the spear dropped from the inside of his hand, and he cried over him'

65. *'masirembiko!'*   *nda ya enana*
    ma-siri-e-mbi-ko   nda ya enana
    ITR.IMP-go.down-come-IM.FUT-2DL   and papa PROX.SG
    mbukamendakəto,   *mbisirekəto,*
    mbu-kame-anda-k-to   mbi-siri-e-k-to
    3.ERG-call.out-3DL.DAT-FR.PAST-DEP   3DL-go.down-come-FR.PAST-DEP
    mbuturaritak   *'sen nor kapu tənasi'*
    mbu-t-rari-ta-k   sen nor kapu t-na-si
    3.ERG-COM-cry-OUT-FR.PAST   son man big PFV-3SG-do

    'you (DL) come down here!' this papa called out to them (DL), and they (DL) came down and he cried '(my) son has become a grown man'

66. *'awo, ma aynde mararaŋənaya.*   *mana nanan*
    awo ma aynde ma-ra-ar-aŋ-naya   ma-na nanan
    yes, 1SG here 1SG-stay-PROG-PRES-1SG   1SG-POSS mama
    yowana tənaona,   *indan kapu yowa ma ŋga nda*
    yowa t-na-o-na-a   indan kapu yowa ma ŋga nda
    DIST PFV-3SG-go-UP-PFV   house big DIST 1SG DAT and
    sik.   *ma nana irormandək ŋatəraraŋənaya*
    si-k   ma nana iror-mandək ŋa-t-ra-ar-aŋ-naya
    make-FR.PAST 1SG mama tree-female INV-COM-stay-PROG-PRES-1SG
    marukorimbiko!'
    ma-rukor-mbi-ko
    ITR.IMP-go.ashore-IM.FUT-2DL

    'yes, I live here. That mama of mine, who built that big house for me, has gone upriver (into the forest). My mama, a tree spirit, looks after me. You (DL) come ashore!'

67. *taməndodə naŋgatək taménd rumondiatətak*
    tamənd-ondə naŋgatik taménd ru-mondi-at-ta-k
    fish-3DL river.current fish shoot-penetrate-?-OUT?-FR.PAST

    tamənd   pisiki   pisiki   mbirakəmbaya
    tamənd   pisiki   pisiki   mbi-ra-k-mbaya
    fish   plenty plenty 3DL-stay-FR.PAST-3DL
    'they (DL) had fish that they (DL) shot on the current, lots and lots of fish'

68. nuŋgo   kapandi   mbuturukorəkə
     nuŋgo   kapa-ndi   mbu-t-rukor-k
     very   big-ADV   3.ERG-CAUS-go.ashore-FR.PAST
     'they (DL) brought ashore a very big (amount of fish)'

69. mayn   ŋga   ramay   ŋgari   mumoran   mbusi
     mayn   ŋga   ramay   ŋgari   mə-moran mbu-si
     husband DAT younger.brother DAT eat-thing 3.ERG-make
     mbupukorata   mbutəmendak
     mbu-pukora-ta mbu-təme-anda-k
     3.ERG-feed-OUT 3ERG-say-3DL.DAT-FR.PAST
     'she (the woman) made food for her husband and the younger brother, fed them and then told them (DL)"

70. 'mana   nana   e   ŋga sikəmb   o   akənjim
     ma-na   nana   e   ŋga si-k-mb   o   akən-jim
     1SG-POSS mama come DAT do-FR.PAST-OBL sun-cloud
     karam ekəmb yowa mamrayn pamba was nda pamba
     karam ekəmb yowa mamrayn pamba was nda pamba
     storm e-k-mb DIST thunder only   wind and only
     yo nanan mana   mbaenaŋok
     yo nanan ma-na   mba-e-n-aŋ-okə
     D   mama 1SG-POSS 3.ERG/1.OBJ-come-LK-PRES-1PC
     'when my mama wants to come, if dark clouds come, with thunder and wind, my mama is coming to us (PC)'

Note the use of the ergative agreement suffix -*okə* 1PC to indicate the object of an otherwise intransitive verb (although semantically the argument is a goal, it cannot be a dative here, as dative suffixes precede tense suffixes and further dative suffixes are not possible with local goals). The usual object suffix for PC -*iya* is also not possible here, because it would normally indicate a PC subject for this verb in normal intransitive use: *e-n-aŋ-iya* come-LK-PRES-PC 'they (PC) are coming'. Note also that this suffix cannot be interpreted as marking the subject, as *mba-* indicates a non-first person subject, either a third person subject or a second person subject by impersonalization.

71. ŋgiramakəto, ŋgirarəkə, nda
    ŋgi-ra-ma-k-to ŋgi-ra-ar-k nda
    3PC-stay-AFTERNOON-FR.PAST 3PC-stay-PROG-FR.PAST and
    wakəna nana e ŋga usianakoya nda
    wakəna nana e ŋga u-si-ana-k-oya nda
    afternoon mama come DAT 3SG-do-3SG.DAT-FR.PAST-3SG and
    akənjim mamrayn nda pamba was nda pamba nda
    akən-jim mamrayn nda pamba was nda pamba nda
    sun-cloud thunder and only wind and only and
    mbutəmendak
    mbu-təme-anda-k
    3.ERG-tell-3DL.DAT-FR.PAST
    'they (PC) stayed until afternoon, and in the afternoon, mama wanted to come and (there were) storm clouds and thunder and wind, and she (the woman) told them (DL)'

72. 'mana nana yowak mbaenandəkək'
    mana nana yowa-okə mba-e-n-andə-kəkə
    ma-na nana DIST-1PC 3.ERG/1.OBJ-come-LK-PFV-1PC
    'we (PC) have my mama, she has come to us (PC)'

73. nda mbətotakəto inda tiŋgi o aynde
    nda mbu-t-o-ta-k-to inda tiŋgi o aynde
    and 3.ERG-CAUS-go-OUT-FR.PAST-DEP house behind here
    mbukakraritak, mayn muna ramay
    mbu-ka-krari-ta-k mayn mu-na ramay
    3.ERG-FIRST-hide-OUT-FR.PAST husband 3SG-POSS younger.brother
    nda
    nda
    and
    'and she took them (DL) out behind the house and hid them (DL) here first, her husband and the younger brother'

74. nanan yo enakəto patukurimba moran
    nana yo e-ana-k-to patukuri-mba moran
    mama DEF come-3SG.DAT-FR.PAST-DEP far.away-LOC thing

mbutundatək
mbu-t-ndat-k
3.ERG-CAUS-know-FR.PAST
'the mama came to her and from a long way she had understood the thing (i.e. what had been going on)'

75. enakəto,   eranak   inda   sur o
    e-ana-k-to   e-ra-ana-k   inda   sur o
    come-3SG.DAT-FR.PAST-DEP come-stay-3SG.DAT-FR.PAST house on
    ambisen   mbutəmek
    ambisen   mbu-təme-k
    daughter 3.ERG-tell-FR.PAST
    'she arrived and came on top of the house, and told her daughter'

76. 'mi,   nor   ombe nakataenande'
    mi   nor   ombe na-kata-e-n-andə-e
    2SG   man   one   3SG-speak-come-LK-PFV-Q
    'you (SG), I think maybe a man has come?'

77. 'kaya,   ke   ɲja   iramanaŋgəbake'
    kay-oya ke   ɲja   i-ra-ma-n-aŋg-mbake
    NEG-1SG 1DL   only   1-stay-DUR-LK-PRES-1DL
    'no, only us (DL) are staying (here)'

78. 'kayna,   nor   ombe   tənaena   minak,
    kay-ona   nor   ombe   t-na-e-n-a   mi-na-k
    NEG-2SG   man   one   PFV-3SG-come-LK-PFV   3SG-POSS-NE
    aramoranomen?,   masinone?,   ramone?,   yanone?,
    ara-moran-omen   masin-ona-e   ram-ona-e   yan-ona-e
    what-thing-wh   brother-2SG-Q younger.sister-2SG-Q papa-2SG-Q
    okumb   o?   mayn   ekəmb   o   kitarəkəna
    okumbi   o   mayn   e-k-mb   o   kit-arək-naya
    PL   husband   come-IRR-OBL   hide-PROHIB-2SG
    mana   ŋga   tawmb   kusamborerətaka
    ma-na   ŋga   taw-mb   ku-sambo-rer-ta-ka
    1SG-POSS   DAT   below-OBL   TR.IMP-put-spread-OUT-2SG
    'no, a man has come, yours (SG). What? Do you (SG) have a brother? Do you (SG) have a younger brother? Do you (SG) have a papa? More than one? If a husband has come, don't hide him from me, bring him out to me'

79. nda   ambisen   nunon   ukaonakəto,
    nda   ambisen   nunon   u-ka-o-ana-k-to
    and   daughter mind   3SG-cut.into-go-3SG.DAT-FR.PAST-DEP
    nda   mbutəmek
    nda   mbu-təme-k
    and   3.ERG-tell-FR.PAST
    'and the daughter thought it over and told her'

80. 'mayn   manak   o   enaŋgoya   muna
    mayn   ma-na-k   o   e-n-aŋg-oya   mu-na
    husband 1SG-POSS-NE   come-LK-PRES-3SG 3SG-POSS
    ram   nda'
    ram   nda
    younger.brother   COM
    'the husband of mine comes with his younger brother'

81. 'ndak   mbamena?
    ndak   mbamena
    where   who.DL
    'where are they (DL) (literally 'who.DL')?'

82. 'embaya   inda   tiŋgi'
    embaya   inda   tiŋgi
    PROX.DL   house   behind
    'they (DL) are behind the house'

83. 'kutukusatae'
    ku-t-kusa-ta-e
    TR.IMP-CAUS-come.out-OUT-Q
    'bring them (DL) out'

84. nda   mbikusamenak
    nda   mbi-kusa-am-e-ana-k
    and   3DL-come.out-DETR-come-3SG.DAT-FR.PAST
    mbutetakəto   inda   pwap
    mbu-t-e-ta-k-to   inda   pwap
    3.ERG-CAUS-come-OUT-FR.PAST-DEP   house   middle
    'and they (DL) came out to her; she (the daughter) brought them to the middle of the house'

85. 'ara   ŋga   imuraŋgime?   tawmb   rama   makondukre?
    ara   ŋga   i-mur-aŋg-iya-ome   taw-mb   ra-ma   mak-onduk-r-e
    what   DAT   2-fear-PRES-PC-wh   below-OBL stay-DUR   bad-FUT-?-Q
    ma   ŋguna   ŋga   prerambiye?   sapikindi   pamba
    ma   ŋgu-na   ŋga   prera-mbi-oya-e   sapik-ndi   pamba
    1SG   2PC-POSS   DAT   angry-IM.FUT-1SG-Q right-ADV   only
    tenanadikəkəko'
    t-e-n-andə-kəko
    PFV-come-LK-PFV-2DL
    'why are you (PC) afraid? Would it be bad to stay below (in clear view)?
    Would I be angry with you (PC)? It's really right that you (DL) have come'

Note the alternation between PC and DL number in this example. Presumably this is because the questions are addressed to the whole group of daughter, son and the two brothers, while it is only the two brothers that have recently arrived.

86. nda   nana   nimbren   okumbi   mbutetanak
    nda   nana   nimbren   okumbi   mbu-t-e-ta-ana-k
    and   mama   pig   PL   3.ERG-CAUS-come-OUT-3SG.DAT-FR.PAST
    'a   masiribiduku'
    a   ma-siri-mbi-onduku
           ITR.IMP-go.down-IM.FUT-2PC
    'and mama brought out her pigs, 'you (PC) go down''

87. nimbren   awr   gimbutuk   ŋgisirambək
    nimbren   awr   ŋgi-mbu-t-k   ŋgi-sira-am-k
    pig   fire   3PC-cook-APPL-FR.PAST 3PC-cut.up-ABOUT-FR.PAST
    mumoran   ŋgisik   ŋgimək
    mə-moran   ŋgi-si-k   ŋgi-mə-k
    eat-thing   3PC-make-FR.PAST   3PC-eat-FR.PAST
    ŋgiramak   ŋgirasek
    ŋgi-ra-ma-k   ŋgi-ra-se-k
    3PC-stay-AFTERNOON-FR.PAST   3PC-stay-NIGHT-FR.PAST
    ŋgikandəksek
    ŋgi-kandək-se-k
    3PC-sleep-NIGHT-FR.PAST
    'they (PC) cooked the pigs in a fire and they (PC) cut them up; they (PC) made food and they (PC) ate it; they (PC) stayed until afternoon, they (PC) stayed until night, and they (PC) slept'

88. *akən   upratanəŋgrak   gibirarək*
    akən   u-pra-ta-anəŋgra-k   ŋgi-mbi-ra-ar-k
    sun   3SG-excrete-OUT-3PC.DAT-FR.PAST 3PC-AGAIN-stay-DAY-FR.PAST
    akən   enamb   nda wakəna   nda nanan
    akən   ena-mb   nda wakəna   nda nanan
    sun   PROX.SG-OBL   and afternoon and mama
    mbutəmekədədə
    mbu-təme-k-ondədə
    3.ERG-tell-FR.PAST-3PC
    'the sun came up, and again they (PC) stayed during the day and on this day in the afternoon, they (PC) told mama (the female tree spirit)'

89. '*ma   nəmand   ŋgatondukoya   sen   nda   mbəna   uyap*
    ma   nəmand   ŋga-t-o-onduk-oya   sen   nda   mbə-na   uyap
    1SG   woman   FIRST-CAUS-go-FUT-1SG   son   COM   3DL-POSS   garden
    o   moran   sitak   o   pate   kapabodək   moran
    o   moran   si-ta-k   o   patek   kapambondə   moran
          thing   make-OUT?-FR.PAST   before long.time?   thing
    mbusamborerukondə   moran   arata   naŋgu   tiŋgi   kandaŋ
    mbu-sambo-rer-k-ondə   moran   arata   naŋgu   tiŋgi   kandaŋ
    3.ERG-put-sow-FR.PAST-3DL   thing   good   skin   behind ?
    ɲja   mbəna   ŋga   enamb   nambrin   nasayndəkoya
    ɲja   mbə-na   ŋga   ena-mb   nambrin   na-say-ndək-oya
    just   3DL-POSS   DAT   PROX.SG-OBL   eye   1SG.ERG-see-NR.PAST-1SG
    '*mbiramanande   e   mbiprenande?*'
    mbi-ra-ma-n-andə-e   e   mbi-pre-n-andə-e
    3DL-stay-DUR-LK-SEQ-Q   or   3DL-die-LK-SEQ-Q
    ndakatotan
    nda-ka-t-o-ta-n
    NOW-FIRST-CAUS-go-OUT-IMP
    The older brother speaks: 'I will take the woman with her son. The things they (DL) grew in their (DL) garden, the food they cultivated a long time ago, all good for them (DL). Of this, I saw: 'are they (DL) still alive or have they (DL) died?' Right now let me take them away (from here)'

90. *nana   katanəŋgrak   'sapikindiko*
    nana   kata-nəŋgra-k   'sapik-ndi-ko
    mama   speak-3PC.DAT-FR.PAST   right-ADV-2DL

maŋgaobiduku   kuŋgatotaka
ma-ŋga-o-bi-onduku   ku-ŋga-t-o-ta-ka
ITR.IMP-FIRST-go-IM.FUT-2PC TR.IMP-FIRST-CAUS-go-OUT-2SG
mambiebiduku   *mana   ŋga   samayt   ŋga*
ma-mbi-e-mbi-onduku   ma-na   ŋga   samayt   ŋga
ITR.IMP-AGAIN-come-IM.FUT-2PC 1SG-POSS DAT see   DAT
nuŋgo   si   obeb   warikiya'
nuŋgo   si   obeb   o-arək-iya
very   do   only   go-PROHIB-PC

'mama said to them (PC): 'you (DL) are right. You (PC) go first, you (SG) take them (DL) away (from here) now. But you (PC) come back to see me, don't go for good'

91. *nanan   yowa   moran   natadaba*
    nanan   yowa   moran   na-tandamba
    mama   DIST   thing   3SG-prepare
    'this mama got things ready'

92. *tumbuna   kapan   urikeranəŋgrak,   moran*
    tumbuna   kapan   u-riker-anəŋra-k   moran
    morning   big   3SG-get.up-3PC.DAT-FR.PAST   thing
    mbutadabanəŋgrak,   *uyap   sur   onəŋrak,*
    mbu-tandamba-anəŋra-k   uyap   sur   o-nəŋra-k
    3.ERG-prepare-3PC.DAT-FR.PAST   garden into   go-3PC.DAT-FR.PAST
    piniman   ndikin   muɲam   manduwaran,   moran
    piniman   ndikin   muɲam   manduwaran   moran
    banana   sweet.potato   sugarcane taro   thing
    mbutukokorasanəŋgrak
    mbu-t-kora-sa-anəŋra-k
    3.ERG-CAUS-gather-IN-3PC.DAT-FR.PAST

    'early morning she gets up for/from them (PC) and prepares things; she goes into the garden for them (PC): bananas, sweet potatoes, sugarcane, taro, she gathered these things for them (PC)'

93. *usiranəŋgrak,*   *kayn   kapu   suman   muna*
    u-sir-anəŋra-k   kayn   kapu   suman   mu-na
    3.SG-go.down-3PC.DAT-FR.PAST   canoe   big   very   3SG-POSS
    kayn   təkararək   *ikun   kayn   yo   okumbi*
    kayn   t-kar-ar-k   ikun   kayn   yo   okumbi
    canoe   CAUS-walk-PROG-FR.PAST   snake   canoe   DEF   PL

namən mbunditanəŋgrak, kayn kapu suman
namən mbu-ndi-t-anəŋgra-k kayn kapu suman
leg 3.ERG-hit-APPL-3PC.DAT-FR.PAST canoe big very
ukusamanəŋgrak
u-kusa-am-anəŋgra-k
3SG-go.out-DETR-3PC.DAT-FR.PAST
'she went down (to the river bank) for them (PC), and she kicked her leg against her canoe for them (PC), (one of) the snake canoes that she traveled in, and it turned into a very big canoe for them (PC)'

94. arəm payndəp ŋginaɲjakəto, moran
    arəm payndəp ŋgi-naɲja-k-to, moran
    river beside 3PC-be.at.bank-FR.PAST-DEP thing
    ŋginiɲjutuk,
    ŋgi-ni-ɲjutu-k
    3PC-put.inside-load.canoe-FR.PAST
    'they (PC) were by the river bank and they (PC) put everything into the canoe'

95. moran ŋginiɲjutuk, nanan ŋgitəmek
    moran ŋgi-ni-ɲjutu-k nanan ŋgi-təme-k
    thing 3PC-put.inside-load.canoe-FR.PAST mama 3PC-tell-FR.PAST
    'they loaded everything inside the canoe, put everything inside the canoe, and they (PC) told mama'

96. 'maŋaramaka paŋgə mandakaokə
    ma-ŋga-ra-ma-ka paŋgə ma-nda-ka-o-okə
    ITR.IMP-FIRST-stay-DUR-2SG 1PC ITR.IMP-NOW-FIRST-go-1PC
    ambiembikə mina ŋga samayt ŋga'
    ma-mbi-e-mbi-okə mi-na ŋga samayt ŋga'
    ITR.IMP-AGAIN-come-IM.FUT-1PC 2SG-POSS DAT see DAT
    'you (SG) stay at first. Let us (PC) go right now, but we (PC) will come back to see you (SG)'

97. 'e, nana, pre ndəkəna mi mana ŋga
    e nana pre ndək-na mi ma-na ŋga
    eh, mama die PURP-1SG 2SG 1SG-POSS DAT
    ŋgaerakotəkənaya mana sen kapu təsi
    ŋga-e-ra-kot-k-naya ma-na sen kapu t-Ø-si
    INV-come-stay-carry-FR.PAST-1SG 1SG-POSS son big PFV-3SG-do

nor   kapu   təsi.   maŋaramaka
nor   kapu   t-Ø-si   ma-ŋa-ra-ma-ka
man   big   PFV-3SG-do   ITR.IMP-FIRST-stay-DUR-2SG
ambiembiya   mina   ŋga   samayt   ŋga
ma-mbi-e-mbi-oya   mi-na   ŋga   samayt   ŋga
ITR.IMP-AGAIN-come-IM.FUT-1SG 2SG-POSS DAT   see   DAT
'eh, mama, I want to die, you (SG) took me in and my son grew up, he become a man. You (SG) stay first and I will come back to see you (SG)'

98. 'nda   'monduk   ratamb   o'
    nda   'ma-o-onduku   rata-mb   o
    and   ITR.IMP-go-2PC now?-OBL
    'you (PC) go now'

99. məŋə   kayn   ŋikandaŋa   ŋikusamə,   nanandək   yo
    məŋə   kayn   ŋi-kanda-ŋa   ŋi-kusa-am   nanandək   yo
    3PC   canoe   3PC-board-go.into 3PC-go.out-DETR   mama   DEF
    rari   nda   uyumanəŋrak   inda   su
    rari   nda   u-yuma-anəŋra-k   inda   sur
    cry   COM   3SG-turn.back-3PC.DAT-FR.PAST   house   into
    eranəmbak
    e-ra-anəmba-k
    come-stay-3DL.DAT-FR.PAST
    'they (PC) board the canoe and depart. The mama crying turned around from them (PC) and went into and stayed in their (DL) house'

100. nda   pamba   rari   nda   ŋgisinek   ngiok
     nda   pamba   rari   nda   ŋi-sina-e-k   ŋi-o-k
     and   only   cry   COM   3PC-go.up-come-FR.PAST   3PC-go-FR.PAST
     ngiokə   məŋəna   numot
     ŋi-o-k   məŋə-na   numot
     3PC-go-FR.PAST   3PC-POSS   village
     'and so while crying they (PC) went upriver, and they (PC) went and they (PC) went to their (PC) village'

101. ŋgirukorə,   moran   mbuturukorəsakədədə
     ŋi-rukor   moran   mbu-t-rukor-sa-k-ondudu
     3PC-go.ashore   thing   3.ERG-CAUS-go.ashore-IN-FR.PAST-3PC

inda   sur   ŋgisamborer   nda   ŋgirasek
inda   sur   ŋgi-sambo-rer   nda   ŋgi-ra-se-k
house   in   3PC-leave-scatter   and   3PC-stay-NIGHT-FR.PAST
ŋgikandəksek
ŋgi-kandək-se-k
3PC-sleep-NIGHT-FR.PAST
'They (PC) went ashore and unloaded everything ashore and deposited that in the house and stayed until nightfall and then they (PC) slept'

102. *gibiramak*   *akən   usisek*
     ŋgi-mbi-ra-ma-k   akən   u-si-se-k
     3PC-AGAIN-stay-AFTERNOON-FR.PAST   sun   3SG-become-NIGHT-FR.PAST
     'they (PC) stayed again through the afternoon and night fell'

103. *mayn   munak   kuŋgam*
     mayn   mu-na-k   kuŋgam
     husband   3SG-POSS-NE spear
     mbuniŋabuk   *kay   sur   o*
     mbu-ni-ŋga-mbu-k   kay   sur   o
     3.ERG-put.inside-go.into-in.canoe-FR.PAST   canoe   into
     'her husband filled up a canoe with his spears'

104. *nda   mayn   umakoya*   *urukorək*
     nda   mayn   u-ma-k-oya   u-rukor-k
     and   husband   3SG-row-FR.PAST-3SG   3SG-go.ashore-FR.PAST
     nanan   yo
     nanan   yo
     mama   DEF
     'and the husband rowed and went ashore where the mama was'

105. *irormandək   nanan   yo   mora   kanda   sik   aymbor*
     iror-mandək   nanan   yo   mora   kanda   si-k   aymbor
     tree-female   mama   DEF   thing   sick   feel-NR.PAST   hearth
     payndəp   kandəksirik   *yowa*
     payndəp   kandək-siri-k   yowa
     beside   sleep-MORNING-FR.PAST   DIST
     'the mama, the female tree spirit, was feeling sick and was sleeping in the morning there by the hearth'

106. urukorukoya   aynəmbep   enamb
     u-rukor-k-oya   aynəmbep   ena-mb
     3SG-go.ashore-FR.PAST-3SG kind.of.spear PROX.SG-OBL
     mbukaməŋjaŋatək
     mbu-kamənjaŋat-k
     3.ERG-stick.into.ground-FR.PAST
     'he came ashore and stuck this *aynəmbep* spear into the ground'

107. 'e, mi   meme?   mana   ambisen ondukoya.   ma   mora
     e,   mi   mi-ome   ma-na   ambisen o-ndək-oya   ma   mora
     e,   2SG   2SG-wh   1SG-POSS daughter go.NR.PAST-3SG   1SG   thing
     kandan   masimanaŋgənaya.   mi   mana   ŋga   bagari
     kandan   ma-si-ma-n-aŋg-naya   mi   ma-na   ŋga   bagari
     sick   1SG-feel-DUR-LK-PRES-1SG 2SG   1SG-POSS   DAT   kill
     ŋga   enaŋgənae?'
     ŋga   e-n-aŋg-na-e
     DAT   come-LK-PRES-2SG-Q
     'eh, who are you? My daughter has gone. I feel sick. You (SG) come to kill me?'

108. mburumondikə,   mbubagarikəto
     mbu-ru-mondi-k   mbu-mbaŋgari-k-to
     3.ERG-shoot-penetrate-FR.PAST 3.ERG-kill-FR.PAST-DEP
     mbutuprek
     mbu-t-pre-k
     3.ERG-CAUS-die-FR.PAST
     'he speared her; he killed her, he killed her'

109. nor   enana   mbutuprekəto,
     nor   enana   mbu-t-pre-k-to
     man   PROX.SG   3.ERG-CAUS-die-FR.PAST-DEP
     marekambuŋ   awr   mbumbu   tamənd   kapu
     mareŋ-kambuŋ   awr   mbu-mbu   tamənd   kapu
     sago-sago.mixed.with.coconut fire 3.ERG-cook fish   big
     ombe   awr   mbumbuana
     ombe   awr   mbu-mbu-ana
     one   fire   3.ERG-cook-3SG.DAT
     'this man killed her and cooked a big fish with sago mixed with coconut for her on the fire'

110. *asumb   sur   mbusurorariana*
     asumb   sur   mbu-sur-o-ra-ri-ana
     mouth   into   3.ERG-inside-go-stay-DOWN-3SG.DAT
     mbutukrotana   *marekambuŋ*   *nda*
     mbu-t-krot-ana   mareŋ-kambuŋ   nda
     3.ERG-CAUS-sew.up-3SG.DAT   sago-sago.mixed.with.coconut   and
     tamənd   nda
     fish   and
     'he jammed the sago mixed with coconut and fish down into her mouth and sewed her up'

111. *mbutukusaməkəto*   *inda   ta,*   *inda-imbot*
     mbu-t-kusa-am-k-to   inda   ta   inda-imbot
     3.ERG-CAUS-go.out-DETR-FR.PAST-DEP house   SOURCE house-nose
     mbuturikerakəto,   *inda-imbot*
     mbu-t-riker-k-to   inda-imbot
     3.ERG-CAUS-get.up-FR.PAST-DEP   house-nose
     mbuturəməkəto,   *mbukənəmbəkəto,*
     mbu-t-rəmə-k-to   mbu-kənəmb-k-to
     3.ERG-CAUS-stand-FR.PAST-DEP   3.ERG-tie-FR.PAST-DEP
     mbusambotək
     mbu-sambo-t-k
     3.ERG-leave-APPL-FR.PAST
     'he took her out from the house, and stood her up on the house gable and tied her and left her (there)'

112. *mu   umbikusamukoya*   *kay   kandaŋgak*
     mu   u-mbi-kusa-am-k-oya   kay   kanda-ŋga-k
     3SG   3SG-AGAIN-go.out-DETR-FR.PAST-3SG   canoe   board-go.into-FR.PAST
     mbiok   *numot*
     mbi-o-k   numot
     AGAIN-go-FR.PAST   village
     'he went out again and boarded his canoe and went back to his village'

113. *okəto*   *numot,   urukorək*
     o-k-to   numot   u-rukor-k
     go-FR.PAST-DEP   village 3SG-go.ashore-FR.PAST
     'he went back to his village and went ashore'

114. saran   ombe   kaynda   kata   muna
     sara-n   ombe   kay-nda   kata   mu-na
     report-NMLZ   one   NEG-3SG   speak   3SG-POSS
     uramak
     u-ra-ma-k
     3SG-stay-AFTERNOON-FR.PAST
     'he said nothing and he stayed through the afternoon'

115. nda   rakamda   nana   ndurum   yowa   onakəto,
     nda   rakamda   nana   ndurum   yowa   o-ana-k-to
     and   night   mama   spirit   DIST   go-3SG.DAT-FR.PAST-DEP
     ambisen   mbutusisek
     ambisen   mbu-t-si-se-k
     daughter   3.ERG-CAUS-feel-NIGHT-FR.PAST
     'and that night that spirit of the mama went to her and stirred her daughter'

116. 'ke   mora   sin   ombe   mbam   o'
     ke   mora   si-n   ombe   mbam   o'
     1DL   thing   do-NMLZ   one   maybe
     'maybe something (happened) to us (DL)'

117. nanan   sen   yo   mbutəmeɲjak
     nanan   sen   yo   mbu-təmeɲja-k
     mama   son   DEF   3.ERG-tell-FR.PAST
     'the mama told the son'

118. 'apasen   minak   o   mora   sin   ombe
     apasen   mi-na-k   o   mora   si-n   ombe
     grandparent   2SG-POSS-NE   thing   do-NMLZ   one
     ukatakusambəmande.   ke   rari   tumbuna
     u-kata-kusa-am-ma-andə-e   ke   rari   tumbuna
     3SG-speak-go.out-DETR-AFTERNOON-PFV-Q   1DL   tomorrow   morning
     kapan   mandaosirike'
     kapan   ma-nda-o-siri-ke
     big   ITR.IMP-NOW-go-go.down-1DL
     'I think something maybe happened to your grandmother. Early tomorrow morning let us (DL) go downriver'

119. rakamda   kapan   mumoran   okombi   nani
     rakamda   kapan   mə-moran   okombi   nani
     night   big   eat-thing   PL   pot
     mbututasirik   piniman   o   rikin   o
     mbu-t-ta-siri-k   piniman   o   rikin   o
     3.ERG-CAUS-boil-go.down-FR.PAST   banana   sweet.potato
     ndurum   rikin
     ndurum   rikin
     spirit   sweet.potato
     'while still dark, she boiled all kinds of food in a pot, banana, sweet potato, spirit sweet potato'

120. akən   ɲja   ukwayramesiranak   sen
     akən   ɲja   u-kwayram-e-siri-ana-k   sen
     sun   only   3-break-come-go.down-3SG.DAT-FR.PAST   son
     mbutoniɲɲjutukoya   mbi-ma-k
     mbu-t-o-ni-ɲjutu-k-oya   mbi-ma-k
     3.ERG-CAUS-go-put.inside-load.canoe-FR.PAST-3SG   3DL-row-FR.PAST
     'just at daybreak, she got her son and loaded the canoe and they (DL) rowed'

121. mbimak   mbimak   mbimakə,   mbənak
     mbi-ma-k   mbi-ma-k   mbi-ma-k   mbə-na-k
     3DL-row-FR.PAST   3DL-row-FR.PAST   3DL-row-FR.PAST   3DL-POSS-NE
     naɲja   rukorikindi,   aynde   inda-imbot   nambrin
     naɲja   rukor-k-ndi   aynde   inda-imbot   nambrin
     river.bank   go.ashore-FR.PAST-ADV   here   house-nose   eye
     urukoranakoya   nanan   makindi
     u-rukor-ana-k-oya   nana   mak-ndi
     3SG-go.ashore-3SG.DAT-FR.PAST-3SG   mama   bad-ADV
     sik   mbusamaytək
     si-k   mbu-samayt-k
     happen-FR.PAST   3.ERG-see-FR.PAST
     'they (DL) rowed and rowed and rowed, and going ashore at their (DL) river bank, they (DL) looked at her on the gable of the house here, and they (DL) saw that mama had turned out badly'

122. e,   nana,   mi   dade   mi   nde   isindəkəname?'
     e   nana   mi   nda-nde   mi   nde   i-si-ndək-na-ome
     eh, mama   2SG and-how   2SG how   2-happen-NR.PAST-2SG-wh
     'hey, mama, what happened to you (SG)?'

123. *mbiraposirikəmbaya*   *muna* *sen* *enana*   *nda*
     mbi-rapo-siri-k-mbaya   muna   sen   enana   nda
     3DL-run-go.down-FR.PAST-3DL   mu-na   son   PROX.SG   and
     nanan   *mbuturariarəkondə*
     nanan   mbu-t-rari-ar-k-ondə
     mama   3.ERG-COM-cry-DAY-FR.PAST-3DL
     'they (DL) ran down and mama and her son cried over her all day'

124. *mbuturarimakondə*   *wakəna*
     mbu-t-rari-ma-k-ondə   wakəna
     3.ERG-COM-cry-AFTERNOON-FR.PAST-3DL   afternoon
     mbukawarikondə
     mbu-kawari-k-ondə
     3.ERG-bury-FR.PAST-3DL
     'they (DL) cried over her until afternoon and in the afternoon they buried her'

125. *nda* *mbirasek*   *mbikandəksek*   *inda*
     nda   mbi-ra-se-k   mbi-kandək-se-k   inda
     and   3DL-stay-NIGHT-FR.PAST 3DL-sleep-NIGHT-FR.PAST   house
     sur   *o*   *akən*   *upratanəmbak*
     sur   o   akən   u-pra-ta-anəmba-k
     inside   sun   3SG-excrete-OUT-3DL.DAT-FR.PAST
     'and they (DL) stayed the night and slept inside the house until dawn broke on them (DL)'

126. *mumoran*   *mbusikondə*   *mbibisiok*
     mə-moran   mbu-si-k-ondə   mbi-mbi-si-o-k
     eat-thing   3DL-make-FR.PAST-3DL   3DL-AGAIN-do-go-FR.PAST
     'they (DL) made food and they went back'

127. *mbiosirik,*   *mbiokə*   *nəma*   *pwap,*
     mbi-o-siri-k   mbi-o-k   nəma   pwap
     3DL-go-go.down-FR.PAST   3DL-go-FR.PAST   mid.river   middle

mayn   yowa   nda   naena,   mbəna   ŋga
mayn   yowa   nda   na-e-na   mbə-na   ŋga
husband DIST   and   3SG-come-UP   3DL-POSS   DAT
nambrin   nambisayk,   mayn   yo   mbəna   ŋga
nambrin   na-mbi-say-k,   mayn   yo   mbə-na   ŋga
eye   3SG-AGAIN-see-FR.PAST   husband DEF   3DL-POSS   DAT
usaynəmbakoya
u-say-anəmba-k-oya
3SG-see-3DL.DAT-FR.PAST-3SG
'they (DL) went, and reached midway and that husband came upon them (DL) and saw them (DL) again'

128. mbə   mumoran   yo   mbuturikerəkondə
     mbə   mə-moran   yo   mbu-t-riker-k-ondə
     3DL   eat-thing   DEF   3.ERG-CAUS-get.up-FR.PAST-3DL
     mbutupapaŋgitətak
     mbu-t-papaŋgit-ta-k
     3.ERG-CAUS-throw.out.(paŋgit- RED)-OUT-FR.PAST
     'they (DL) threw out the food they (DL) had prepared'

129. inaŋ   mbusuroramək,   kayn
     inaŋ   mbu-sur-o-ra-am-k   kayn
     oar   3.ERG-inside-go-stay-ABOUT-FR.PAST   canoe
     mbusuroramək,   wanin
     mbu-sur-o-ra-am-k   wanin
     3.ERG-inside-go-stay-ABOUT-FR.PAST   crocodile
     usianak,   inaŋ   yo   uruma   usianak
     u-si-ana-k   inaŋ   yo   uruma   u-si-ana-k
     3SG-do-3SG.DAT-FR.PAST   oar   DEF   shark   3SG-do-3SG.DAT-FR.PAST
     'they (DL) tossed the oar overboard and set the canoe adrift; it turned into a crocodile, the oar turned into a shark'

130. nana   nda   sen   nda   mbipaneŋaramәkəto
     nana   nda   sen   and   mbi-pane-ŋa-ra-am-k-to
     mama   and   son   nda   3DL-jump-go.into-stay-ABOUT-FR.PAST-DEP
     arəm   su,   mayn   mbu-təme-k
     arəm   su   mayn   mbu-təme-k
     river into   husband   3.ERG-tell-FR.PAST
     'mama and son jumped into the river, and she told her husband'

131. *'mi   mana   nanan ibagarindukona,   ma   mina   nda*
     mi   ma-na   nanan i-mbaŋari-ndək-ona ma mi-na   nda
     2SG 1SG-POSS mama 2-kill-NR.PAST-2SG   1SG 2SG-POSS COM
     ra   ndək   tayn   kaya   mana   nanan təpre
     ra   ndək   tayn   kay-oya   ma-na   nanan t-Ø-pre
     stay PURP ability NEG-1SG 1SG-POSS mama PFV-3SG-die
     maprembiya   mi   mambioka   mina
     ma-pre-mbi-oya   mi   ma-mbi-o-ka   mi-na
     ITR.IMP-die-IM.FUT-1SG 2SG   ITR.IMP-AGAIN-go-2SG 2SG-POSS
     numot'
     numot
     village
     'you (SG) killed my mama. I can't stay with you (SG)'. My mama has died; let me die. You (SG) go back to your (SG) village'

132. *nanan nda   sen   nda   pat   mbisik*
     nanan nda   sen   nda   pat   mbi-si-k
     mama   and   son   and   stone 3DL-become-FR.PAST
     'the mama and son turned into stone'

133. *dada   kayn*
     ndanda kay-onda
     more   NEG-3SG
     'it has no more' = 'it's finished'

References

Abbott, Stan. 1985. Nor-Pondo lexicostatistical survey. *Pacific Linguistics* A63.313–338.
Baker, Mark C. 1988. *Incorporation: A Theory of Grammatical Function Changing*. Chicago: University of Chicago Press.
Baker, Mark C. 1996. *The Polysynthesis Parameter*. New York: Oxford University Press.
Bruce, Les. 1984. The Alamblak language of Papua New Guinea (East Sepik). *Pacific Linguistics* C81.
Claas, Ulrike and Paul Roscoe. 2009. A journey up the Sepik River in 1887. *The Journal of Pacific History* 44. 333–343.
Comrie, Bernard. 1974. Causatives and universal grammar. *Transactions of the Philological Society* 73. 1–32.
Comrie, Bernard. 1985. Causative verb formation and other verb-deriving morphology. In Timothy Shopen (ed.), *Language Typology and Syntactic Description: Grammatical Categories and the Lexicon,* 309–348. Cambridge: Cambridge University Press.
Cruikshank, Amy. 2011. *Complementation in South Band Pawnee*. Unpublished Honours thesis, University of Sydney.
Dahlstrom, Amy. 1991. *Plains Cree Morphosyntax*. New York: Garland Publications.
Davidson, Matthew. 2002. *Studies in Southern Wakashan (Nootkan) grammar*. Unpublished PhD dissertation, University of Buffalo.
de Vries, Lourens. 2005. Toward a typology of tail-head linkage. *Studies in Language* 29. 363–384.
Dixon, R. M. W. 1979. Ergativity. *Language* 55. 59–138.
Dixon, R. M. W. 1994. *Ergativity*. Cambridge: Cambridge University Press.
Dixon, R. M. W. 2010. *Basic Linguistic Theory, Volume Two: Grammatical Topics*. Oxford: Oxford University Press.
Dowty, David. 1979. *Word Meaning and Montague Grammar*. Dordrecht: Reidel Publications.
Evans, Nicholas. 2007. Insubordination and its uses. In Irina Nikolaeva (ed.), *Finiteness: Theoretical and Empirical Foundations,* 366–431. Oxford: Oxford University Press.
Foley, William A. 1991. *The Yimas Language of New Guinea*. Stanford: Stanford University Press.
Foley, William A. 2010. Events and serial verb constructions. In Mengistu Amberber, Brett Baker and Mark Harvey (eds.), *Complex Predicates: Cross-linguistic Perspectives on Event Structure,* 79–109. Cambridge: Cambridge University Press.
Foley, William A. 2016. Direct versus inverse in Murik-Kopar. In Jens Fleischhauer, Anja Latrouite and Rainer Osswald (eds.), *Explorations of the Syntax-semantics Interface,* 265–288. Düsseldorf: Düsseldorf University Press.
Foley, William A. 2017a. The languages of the Sepik-Ramu basin and environs. In Bill Palmer (ed.), *The Languages and Linguistics of the New Guinea Area: A Comprehensive Guide,* 177–412. Berlin: de Gruyter.
Foley, William A. 2017b. Polysynthesis in New Guinea. In Michael Fortescue, Marianne Mithun and Nicholas Evans (eds.), *The Oxford Handbook of Polysynthesis,* 336–359. Oxford: Oxford University Press.
Foley, William A. 2017c. Yimas: The profile of a polysynthetic language of New Guinea. In Michael Fortescue, Marianne Mithun and Nicholas Evans (eds.), *The Oxford Handbook of Polysynthesis,* 808–829. Oxford: Oxford University Press.
Foley, William A. 2022. Number in the languages of the Lower Sepik family. In Paolo Acquaviva and Michael Daniel (eds.), *The Handbook of Number,* 529–576. Berlin: de Gruyter.

Foley, William A. and Robert D Van Valin Jr. 1984. *Functional Syntax and Universal Grammar*. Cambridge: Cambridge University Press.
Guillaume, Antoine and Harold Koch (eds.). 2021. *Associated Motion*. Berlin: de Gruyter.
Hale, Kenneth. 1976. The adjoined relative clause in Australia. In R. M. W. Dixon (ed.), *Grammatical Categories in Australian Languages*, 76–105. Canberra: Australian Institute of Aboriginal Studies.
Heath, Jeffrey. 1997. Pragmatic skewing in 1⟷2 pronominal-affix paradigms. *International Journal of American Linguistics* 64. 83–104.
Hockett, Charles. 1966. What Algonquian is really like. *International Journal of American Linguistics* 32.59–73.
Jacques, Guillaume and Anton Antonov. 2014. Direct/inverse systems. *Language and Linguistic Compass* 8. 301–318.
Lipset, David. 1997. *Mangrove Man: Dialogics of Culture in the Sepik Estuary*. Cambridge: Cambridge University Press.
Mattissen, Johanna. 2003. *Dependent-head Synthesis in Nivkh: A Contribution to a Typology of Polysynthesis*. Amsterdam: Benjamins.
Mithun, Marianne. 1984. How to avoid subordination. *Proceedings of the Tenth Annual Meeting of the Berkeley Linguistics Society*, 493–509.
Nichols, Johanna. 1986. Head marking and dependent marking languages. *Language* 62. 56–119.
Nichols, Johanna. 2017. Polysynthesis and head marking. In Michael Fortescue, Marianne Mithun and Nicholas Evans (eds.), *The Oxford Handbook of Polysynthesis*, 59–69. Oxford: Oxford University Press.
Parks, Douglas. 1976. *A Grammar of Pawnee*. New York: Garland Publications.
Pawley, Andrew. 1966. *The Structure of Karam*. Unpublished PhD dissertation, University of Auckland.
Schmidt, Joseph. 1922–1923. Die Ethnographie der Nor-Papua (Murik-Kaup-Karau) bei Dallmanhafen, Neu-Guinea. *Anthropos* 18–19. 700–732.
Schmidt, Joseph. 1926. Die Ethnographie der Nor-Papua (Murik-Kaup-Karau) bei Dallmanhafen, Neu-Guinea. *Anthropos* 21. 38–71.
Schmidt, Joseph. 1933. Neue Beiträge zur Ethnographie der Nor-Papua (Neu-Guinea). *Anthropos* 28. 321–354, 663–682.
Schmidt, Joseph. 1953. *Vokabular und Grammatik der Murik-Sprache in Nordost-Neuguinea*. (Micro-Bibliotheca Anthropos 3). Posieux: Anthropos Institute.
Silverstein, Michael. 1976. Hierarchy of features and ergativity. In R. M. W. Dixon (ed.), *Grammatical Categories in Australian Languages*, 112–171. Canberra: Australian Institute of Aboriginal Studies.
Wolfart, H. Christoph. 1973. Plains Cree: A grammatical study. *Transactions of the American Philosphical Society* 63.5.
Wolvengrey, Arok Elessar. 2011. *Semantic and Pragmatic Functions in Plains Cree Syntax*. Utrecht: LOT Publications.
Woodbury, Hanni. 2018. *A Reference Grammar of the Onondaga Language*. Toronto: University of Toronto Press.
Vendler, Zeno. 1967. *Philosophy in Linguistics*. Ithaca: Cornell University Press.

Index

/ə/ epenethesis 15, 18
/n/ insertion 20, 100–101

ability 109
accomplishments 101
accusative 65–71, 87, 97, 102–103
adjectival verbs 32–34, 36, 97, 175
adjectives 24, 30, 33, 35, 55, 60
adverbials XI, 5, 35, 33, 35, 38, 104, 110, 121, 136, 138–142, 146, 180, 189, 203
adverbial clauses 171, 187
afterthoughts 153–154
agreement 5, 14, 20–21, 24, 28–30, 32–34, 43, 68, 81, 85–86, 90–92, 95, 111, 114, 118, 129, 132–133, 138, 146, 151, 155, 158, 161–162, 175, 178, 194, 226
Algonkian languages 29, 83
allophones 6
Animacy Hierarchy 5, 27, 29, 65, 75, 81, 83, 155, 157
argument bleeding 189, 202–203
aspect 96, 104–105, 136, 143
associated motion 146–147
attribution 175

benefactive 94, 164
bound pronominals XVII, XVII, 39, 66, 68–69, 71–72, 74–75, 77, 79, 83, 85–88, 97, 99–103, 107–109, 112, 115, 118–120, 136, 140, 155, 175–179, 182, 187, 194

case marking 5, 25, 151, 154
causal 172, 193
causative XI, 34, 65, 77, 124–129, 138, 145, 158, 160, 193
clause chaining 5, 34, 52, 101, 180, 187, 189, 191, 193–195, 202–203
codas 10, 11, 15
cognate objects 67–68, 120, 156, 162
comitative applicative 51, 56, 126–129, 138, 170
comitative postposition XI, 36–37, 46–48, 51, 56, 127, 170
conditional clauses 187–188

conjunctions 24, 51–52
consonant clusters 7, 10–11, 14–18, 140
coordination 5, 146, 180, 187, 190, 193
core arguments 4, 5, 29, 65–66, 68, 71, 74–75, 79, 120–123, 125, 127, 133, 136, 151–152, 154–155, 157–158, 160–161, 180, 194
counterfactuals XI, 187–189

dative infinitives 180–181, 183–184
dative postposition 25, 28, 47, 66, 76, 78, 80, 109, 117, 123, 158, 161–163, 177, 180, 183, 184
dative suffixes XI, 25, 47, 66, 77–78, 83, 90–92, 94–95, 129, 132–133, 143, 155, 157, 159, 160–163, 178, 226
day counters 49
definite determiner 43, 61–62
deictics 43, 45, 56, 60, 62
denasalization 20, 21–22, 27, 36, 38, 69, 72, 74, 102, 140, 142, 166–167, 170, 173, 185, 208
dependent verbs 34, 52, 73, 97, 101, 193–194, 196, 199, 202
dependent-head synthesis 4, 135
desideratives XI, 68, 109, 117, 165, 181–183
desyllabification 17, 18–19
detransitivizer XI, 123, 136, 143
direct inflection 5, 29, 38, 65, 71, 75, 77–78, 80, 82–86, 89–90, 92, 123, 157, 160–161
direct-inverse 5, 29, 65, 71, 75, 86, 157
directionals 5, 135, 143
disjunction 193
distal deictic XI, 42–43, 45
ditransitive verbs 66, 76, 78, 80, 83–84, 92, 125, 154, 157, 160–161
durative aspect XI, 73, 105

ergative XI, XVII, 28, 29, 65–68, 71, 74–81, 83, 85–87, 89–90, 97, 99, 100–101, 119, 176, 226
experiential clauses 31, 162–163
extended aspect 106

https://doi.org/10.1515/9783110791440-011

far past tense XI, 24, 31–32, 60, 71, 86, 97, 99, 102, 187
finite verbs 5, 26, 97, 180, 187, 189
focus position 96, 152
future tense XI, 21, 26–27, 49, 68, 97–98, 103–104, 107, 185

general applicative XI, 127, 129–130, 132, 136, 143
genitive suffix 46–47, 57

habitual events 99
head marking 4, 25–26, 39, 151, 154
headless relative clauses 64
heavy noun phrases 153
hortatives 119, 139

identification 151, 173, 175
idioms 110, 145, 157, 164–165
illocutionary force 97, 112
immediate future tense XI, 20, 27, 66, 68–69, 97, 100, 102–103, 107, 110, 115
imperatives XI, XVII, 29–30, 33, 68, 103, 110, 115, 117, 119–120, 122, 124, 139, 141
impersonalization 83–85, 92, 173, 226
impersonal verbs 162
incorporation 4, 5, 104, 122, 133, 135–136, 138, 143, 145
indefinite determiner 62
independent verbs 180, 190–191, 193, 196, 199
infinitives 47, 164, 180, 182
instruments 171–172
insubordination 201
intensifiers 32
interjections 53
interrogative pronouns 40
intransitive verbs XI, XVII, XVII, 13, 21, 24, 28–29, 30, 65–68, 70, 71– 73, 78–79, 83, 86, 91, 115, 118, 120, 121–125, 131–133, 135–136, 151, 154, 156–158, 161–162, 226
inverse inflection XI, 18, 75, 79–84, 89–90, 117, 156, 158, 160–161
irrealis XI, 32, 97, 103, 187

Kalam 8
Kanda 3, 25, 37, 54

local persons 40, 71–72, 75, 77–81, 83, 89–90, 92, 94–95, 117, 158, 160–161
locations 4, 48, 146, 168–170, 172, 174
locative XI, 51, 55, 168
Lower Sepik family XIII, XV, 1–3, 7, 24, 29, 39, 54, 71, 127

malefactive 96, 164
Mambuwan 39
modality XI, 97, 107, 111
mood XI, XVII, 29, 33, 65–68, 96–97, 112, 115, 178, 193
Murik 1, 3, 24, 37, 40–41, 43, 54–56, 80, 140, 205–210

near past tense XI, 49–51, 55–66, 68, 71, 73, 86, 97, 99, 104, 140, 153
necessity 110
negation XI, 30, 68, 97, 107–109, 117, 179
neutral inflection 75, 77–78, 80, 89, 160
Nivkh 135
nominalizations 17, 25–26, 55, 109, 180, 186
nominative XVII, 67, 71, 76, 78–79, 81, 83–84, 89, 118, 177, 189
non-finite 5, 180
non-local person 71–72, 75, 77–80, 90, 92, 94, 157, 159, 160–161
noun compounds 57
noun incorporation 5, 136, 145
noun phrases 25, 57, 59, 154, 178
noun pluralization 56, 60
noun-noun compounds 31
nouns 5, 24–26, 31, 42, 46, 49–51, 54–58, 60, 167, 180
numerals 36, 37, 42, 56, 60, 62
Nuu Chah Nulth 135

objects XI, 24–29, 65, 68–69, 71, 74–76, 78–85, 89–92, 117, –123, 127, 129–134, 151–153, 155, 158, 160, 173, 176, 203, 226
obligation 97, 110, 115

obliques XI, 63, 76, 117, 123, 135, 151, 155, 161, 163, 168, 171, 187–188
Onondaga 135, 136

Pawnee 135
perfective aspect XVII, 7, 19, 27, 71, 86–87, 89–90, 97, 100–102, 104–105, 163, 199
permissive 119
phonemes 6
polypersonalism 4
polysynthesis V, 65, 104, 133, 135–136, 138, 191
possession XVII, 46, 57–60, 64, 95, 109, 133, 176, 178–179
possessive suffix 47, 121, 164
possessor raising 92, 95–96, 133–135, 146, 178
postpositions 7, 17, 25, 41, 46, 48–49, 54–55, 104, 109, 151, 155, 163, 166–168, 170, 180, 188
prenasalized voiced stops 6, 9–11, 15, 20, 22, 86, 69, 72, 140, 166, 208
present tense 4, 21, 26, 43, 60–61, 71–74, 81, 83, 86, 97–100, 105, 125, 140–141, 157
progressive aspect 100, 104
prohibitives 30, 68, 111, 117, 119
pronouns XVII, 39–41, 46, 54, 56–58, 82–83, 85, 121
prosodic particle *o* 167, 185, 187
proximal deictic 41, 43
purposive infinitives 166, 185
purposive postposition XI, 48, 55, 104, 109, 166–167, 180, 185

quantifiers 24, 37, 56

realized events XVII, 6–7, 15, 20, 30, 66–67, 71–72, 74, 83, 86, 97–99, 101, 107–108, 118, 121, 129, 136, 140, 176–177, 180, 182, 189
recipients 25, 65–66, 76, 78, 80, 83–85, 92, 123, 153–154, 157–161, 163, 165
reflexivization 65, 121–122, 138
relative clauses 60–64
relativized noun 61–63

sequential 146–148, 199, 201
serial verb constructions 122, 127, 145–149, 202–203
simultaneous 146–147, 189, 203
Singrin 1–2, 4
source postposition 167
stimulus 129, 162
stress 13
subjects XVII, XVII, 14, 20–22, 24–34, 61, 65–71, 73–85, 87, 90–92, 99, 102–103, 110, 115, 119, 121–123, 125, 127, 129, 133, 135–136, 142, 146, 151–156, 158, 162–163, 173–176, 178, 186, 191, 194, 226
subordinate clauses 180, 187, 189
switch reference 194–195
syllable breaks 11

tail-head linkage 196
temporals XVII, XV, 24, 32, 49–51, 55, 98–99, 135–137, 139, 140, 143, 146, 151, 153
Tok Pisin V, 8, 37, 40, 52, 106, 109, 122, 127, 131, 140, 158, 160–163, 171, 180, 183, 186
transitive verbs 2, XVII, 28–30, 65–71, 75–76, 83, 85, 87, 90–92, 115, 118–125, 129, 131–132, 134–136, 151, 154–155, 158
transitivity 27–28, 66, 68, 91–92, 120
Trans New Guinea family 191, 196

unaccusative verbs 133, 135
uncontrolled events 94, 132
unrealized events XVII, 20, 30, 68–69, 72, 97–98, 102–103, 107, 109, 115, 118–119, 179, 181–183, 187, 189

valence 92, 120, 161
verb incorporation 5, 136, 145–149
verb stems 104, 120–123, 125, 132, 136, 138, 143, 161
verb themes 78, 80, 83–85, 132, 146
voiced stops 6, 8, 10–12, 20–21, 69, 74, 86, 140
vowel harmony 13–18, 55, 57, 99, 104, 121, 185
vowel phonemes XV, 7, 8

Watam 2, 4–5, 22, 37, 43
wh formative 40–41, 113–114
wh-questions 113
word final codas 9
word order 5, 46, 60, 93, 108, 131, 146, 151–153, 155–156, 165, 181, 201

Yes-no questions 112
Yimas 2, 4–5, 8, 11, 13, 24–25, 30, 31, 36–37, 40–41, 49, 54, 97–98, 104, 111, 125, 127, 132, 135, 143, 145–146, 157, 191

www.ingramcontent.com/pod-product-compliance
Lightning Source LLC
Chambersburg PA
CBHW031424150426
43191CB00006B/389